BRAHM

THE KINETIC THEORY OF NATURE

HASMUKH RATHOD

Dedication

This book is Dedicated to
my
late father

"એ અજર છે, એ અમર છે, અકળ છે, નિરાકાર છે,
એને કોઈ રંગ નથી, કોઈ ગંધ નથી,
એ સર્વ-વ્યાપી છે,
એ બધા માં છે અને બધું એમાં છે,
સૌથી પહેલા એ છે અને સૌથી પછી પણ એ છે,
એ ત્યારે પણ હતો જ્યારે કશું નત્,
એ ત્યારે પણ હશે જયારે કશું નહીં હોય,
.....

Einstein said,

"If you can't explain

a theory to a child,

the theory is

probably useless."

Michio Kaku elaborates on it,

"Meaning that great theories are based on simple principles, concepts that you can see visually like pictures. Useless theories are just bunches of algebra that go nowhere. Take a look at Newton's Laws are based on billiard balls, planets, stars going around each other. Very pictorial Newton's laws. So, great theories have a simple pictorial representation."

Self-introduction by the Author

My Name is Hasmukh Devidas Rathod. I am an educated man aged 44 years in 2025. I am an Indian citizen living and working in India. I have a bachelor's degree in medical science and no degree in physics. I am interested in understanding the Nature, and these are some extractions as a physics model from my understanding of the Nature.

I thanks to the Internet, which really is an ocean of information, which helped me a lot. Reading stuffs on internet and watching You-tube videos helped me to refine my imaginations.

If you want to know more about me then a chapter 16. My journey to the Aether is a whole chapter which provides lot of information about me.

Dr. Hasmukh Devidas Rathod.

Disclaimer

The theory represented in this book is my creative original work, but it is related to the whole reality. The reality is one, so some part of it may have similarity with part of some other models or theories. This is because the reality is one and failed model may also contain facts. Author also believes that if someone think enough, he/she must realise that empty space is not empty and there must be something there which is responsible for gravity and other forces. So, similar theories might be already offered by some authors which could not get enough attention due to their incompleteness. Aether is an old topic and there are uncounted opinions by several peoples about it. I am not the only one who believes in Aether. The Indian words like "Brahm, Brahmand, etc" used here are in a use in India since ancient time. Those words are inspired from descriptions in literature of the ancient India, but I explained it in detail in my way which is my imagination or calculations. I used those words here because I think that those should have real meanings as described here. Indian ancient texts are in Vedic Sanskrit which is not fully decoded and have not still completely understood. So, talking about relevance or non-relevance to those texts is not appropriate. Still the theory in this book has close relation with the philosophy of Brahm in Advait Vedanta. Physical world as a manifestation in Brahm due to Maya is very well explained in this book in this theory of Existence. Due to such close relation, I have continued using the world Brahm in this theory. I also found much relevance with mythological religious paintings and stories in Hinduism which I have mentioned in a "chapter 18" of the second part of this book. The Nature of the phonetics pronunciation of the word Brahm gives a feel of entity full of energy leaving resonance as an effect of vibration, that convince me to continue using that word instead of dynamic Aether. So, choice of the words has reasons that I find fitted with the concepts. The meaning of these words described in this book may not have relation with literal or casual meaning of those words in the Indian languages. If the claims or opinion about fundamentals are in support or contradiction with any religious belief, then it remains always there when we are talking about the science, but the aim of representing such a view is not there to hurt feelings of any theist. This book is here to find relevance with religious concepts and not against them.

This theory explains physical world (Srushti) and it doesn't say anything about conscious things including God or Ishwar. When there are many different opinions, and someone provide his own opinion, conflicts with other opinions may be there. If views in this book supports some ideas of the particular one or more religions, then it doesn't mean that these concepts are totally derived from them. This theory only includes concepts that fits in its framework. When I found relation of concepts of this theory with any things then I may have showed it in this book. The book is about reality, and everything is about the real existences. No one including me should have claim for being the first one who have introduced Aether or Brahm very first time, because Aether or Brahm is an ancient concept and well mentioned in various ways in Indian literature and elsewhere also. This theory provides a unique, deeper and clearer insight by describing "Brahm" in a scientific way with detailed logical explanations. This theory not only say, what a Brahm could be, but it also explains how it could be.

This book has tried to explain everything with mechanical sense, so theories like Mechanical push gravity, Vortex theory, Field theory, Electromagnetic theory, some part of relativity...etc" may get woven in this framework. I don't have claim for every part of this theory as my original idea, but the framework with explanations represented here as a whole model is my original work including many original ideas within it. The framework of Brahm that ultimately includes all other things is my original work. Undisputedly advancement in a science guided me to this theory and Ancient Indian Hindu philosophies provided a confidence to believe in this theory.

Contents

Introduction

The word Brahm is taken from the philosophy of the supreme aether like existence that harbours each and every existence within it as a physical world which is its manifested form powered by "Maya". This description of Brahm is there in a Hindu ancient philosophy known as Advait Vedanta means a philosophy of non-dualism. The description of Brahm as a theory of dynamic aether or a kinetic theory of existence is a detailed explanation of Brahm. Here Theory of Brahm don't include the spiritual aspect of the Advait Vedanta, but this theory just explains a Nature with natural mechanical laws which forms a scientific theory of Brahm. This theory of Brahm is described as pure scientific fundamental physics model and nor as a spirituality or an imagination without explanations with laws of physics. Part 2 of this book explains the theory of Brahm.

This book is about "Brahm" which includes an all-pervading multi-scaled Aethers and moving energy carriers in all directions. A key feature that may give you an insight to understand Brahm model, must be a shadow gravity or mechanical push gravity environment that is the expression of This Brahm at every point in the space. It simply means that gravity is not described as an intrinsic property of the mass, but it is a result of the random motions and interaction of moving energy carriers (flux) with the masses. Imagining a mass without an internal property of gravity or any other fundamental attractive force placed in a space without an enough push gravity, then inquiring its fate is the main thought experiment shown in the of the second chapter in a part two of this book which provides most of the insight about this model. Believing a gravity as an intrinsic property of the mass leads to the modern physics, either to the Newtonian physics, General relativity or to the Standard model describing gravitons. Though theory of Brahm agrees with space-time curvature of the general relativity still it differs from general relativity by including Aether and Brahm in a space as well as everywhere. This theory of Brahm don't agree with 1. the calculations based on Lorentz transformation with accepting interchangeability of subjects and 2. calculations of time dilation and length contraction. General relativity includes ability of the mass to curve spacetime matrix while this theory separates gravity from mass and that gravity is described as an effect due to surrounding pushing flux. This Kinetic theory of gravity is explained here with the shadow gravity mechanism, but shadow gravity or mechanical push gravity don't provide such ability to mass Instead, it states that mass neither has intrinsic property of attracting other masses, nor has an ability to curve surrounding by itself alone directly. Role of the mass is passive like a resistance or obstacle in the push gravity environment. When we apply this push gravity mechanism and think about infinite masses in a blank space lacking enough push gravity to hold them intact and try to understand the consequences, where it settles down to a stable steady state it represents a "Brahm" described in this book. This theory shows that mass has tendency to disperse in surrounding, if it is not stabilized by surrounding enough push. Those pressure forming moving energy carriers as masses or waves which provides stability to bigger masses are termed together as "Flux" which is similar to flow of moving corpuscles in gravitational environment in the kinetic theory of gravity. Still the difference is multi scaled flow which forms multi scaled Aethers and Brahm. Other fundamental forces are also described as manifestations of mechanics in this gravitational environment. So, this model doesn't accept any fundamental forces as fundamental, instead, these forces are described as consequences of mechanical interactions. So, this was a brief overview.

Both terms Brahm and Aether ultimately represents the same thing but from different point of view in an infinite matrix. Brahm represents more moving momentum carriers and includes Aether as a manifestation within that environment, while Aether represents a matrix substance Aether that may include vortices of waves within it as a momentum carrier that ultimately creates an environment that could manifest physical world and a push gravity forming environment within it which is described as a "Brahm" in this book. So, basically there is a scaled matrix layers which are termed as Aethers and Vortices of waves within them forming scaled push gravity environments. There all together termed as Brahm. If we apply the term Aether to the possible lowermost self-stable Matrix and describes all other aethereal layers and dynamics of flux as a manifestation due to wave kinetics in this Aether, then everything can be described as this Aether and manifestation within this Aether. But for that we have to accept an undefined Aether as a primary existence without further insight about its stability and its structure. But if we want to inquire the structure of that primary Aether then we can explore it only as a structure created in a Brahm due to kinetics of masses and flux at some scale. So, both terms Brahm and Aether represents one single existing thing called "Nature". If you select Aether as a base than Brahm is manifestation and if you select Brahm as a base than Aether is a manifestation. Physical world is the manifestation in this Aether and Brahm as a wave dynamic. Nature has a large spectrum based on the sizes of the units. Nature includes Brahm at various scales which includes layers of Aether also. If you are in search of an absolute supreme existence here, then this book may disappoint you. This book describes Nature with infinity and no final boundaries. This model doesn't provide an answer to the questions which don't fits to the nature of the Nature. The example of such question is, "what is the smallest thing?" You may find more such questions in a chapter 15, about residual quarries.

The term Nature includes the Whole existence including Brahm and Aether. Nature also includes a physical world as a Manifestation within Brahm, but presence of Physical world is not always necessary because cyclic changes at aethereal scales may include formation and annihilations of physical worlds at every aethereal scales of Brahm. So, Nature includes he existence that may or may not include the physical world but includes Brahm. So, Brahm is the Nature. "Srushti" is the word used specifically for the word Physical world. Srushti is a Manifestation in Brahm as a physical world which can be considered as a "creation" in the sense of religious thinking. Here Srushti is described as a cyclic manifestation within Brahm as a result of natural continuous prosses and not depending on any creator or anything having will power. So, this theory is a self-stable theory of the Nature which includes continuous cycles due to instability in Nature.

Aim of this book is to provide a model of the fundamental physics that must explains most of the observational findings of science. It is the framework that explains different existences and phenomena within it that can provides a way to think. Aim is to provide a model with such a depth that can provide further insight about the Nature which can explain basic laws of the Nature. This model shows simpler framework with mechanical sense at everywhere. There is nothing assumed as a miraculous or mysterious. Most of the concepts used are demystifying and realistic. This is the most possible no miracle realistic approach carried out to understand the Nature. All is dynamics only but to the level that touches infinities.

This Brahm model shows a top view to the Nature as a whole. Here I am using the word "Nature" instead of the word "Universe" because existence is not limited to the Universe only and Nature word represents the entire Existence. This model of the Nature provides a single framework for the physics that can show proper places for possibly all known data without contradicting to one another. This doesn't mean that this model or framework answers all questions or solves all queries, but this just provide a new understanding of Nature and its phenomena that arranges data with sound logic and co-ordinations with one another wherever we can understand the relations. Means it has a one unified view for every existence and phenomena of Nature. This is a logic-based framework with assumptions in the model which are unproved. At some places there may be a gap in a picture which shows unexplored portions of this model, so this is neither the complete model of the Nature or the Universe nor the theory of everything that explains and demystify everything. Still this model has a potential to include everything within itself and this model explains all fundamental forces as a mechanical interaction that preserves laws of classical mechanics. It explains time dilation within boundaries of classical physics without contradicting with it. It also provides space where further advancement may include missing data. *Simply this theory is the extension of the classical physics including modified relativity.* This is just a theory of Existence covering almost everything with some unexplored areas. This model reestablishes a natural thinking toward the Nature. To accept this model or not is the choice of the reader because author is not providing any new experimental proof here. Here there is just an opinion or view and logical framework by the author about the Nature of the existence mostly in a layman language. This is an enhanced old way of thinking about Nature. In this era of the modern physics, is there any scope for the model which don't agree to many leading theories of the mainstream science! Of course, there is not just a scope, but it is the demand of the modern science also, and that's why this theory is here. There are lots of contradictions between many leading theories of the modern physics which don't have a single unified framework. Making mistakes and continue trying is the way of exploration of the Nature. Modern theories don't obey the boundaries of reality and continue to expand in areas of fantasy. No strict law is defined to demark the margin between reality and fantasy. Math and equations are believed as a tool to confirm predictions. No method is available that can detect errors of logic while using maths in modern physics. This all raised a demand for unified framework. This model doesn't allow any imaginary thing or outcome that can't be conceptualise with a realistic mechanical sense. Surprisingly this model opens the door toward mysterious possibilities of the supernatural things as unexplored poorly expressed natural things. Thus, this model actually extends boundary of the physical world toward hidden areas simultaneously closes a door for baseless fantasies.

This book is going to introduce Brahm theory with descriptions of certain characters of Brahm and Aether. Aether described here is not same as luminiferous Aether, Lorentz's electron Aether or any other Aether described previously in a science. This Aether is described as a supreme existence and medium for everything with its complex structure and Nature. Its state is not any of that of the matter, means it is not a gas, fluid, solid, plasma or any known state of the matter. It is a complex thing that has certain specific characters which are well defined in relevant chapters. Sometimes it is denoted as fluid like Aether, but that doesn't confirm it as a fluid. It is a Matrix. It may have symmetry like a relativistic neo-ether that can explain reason behind time dilation.

On the other way if we start from absolute solid mass and absolute blank, then that would lead to Brahm. Which is sufficient to harbor physical world without Aether by accepting light as a particle. This is a simpler model than Brahm-Aether model. But if we want waves, light as a wave then Aether can be a manifestations within this Brahm. In a whole book, a model of Brahm with Aether is described on priority, but one last chapter (chapter 20) is dedicated to the model of Brahm without Aether also.

The concept of time dilation can be explained very well with Brahm model including Aether in chapter 19.

The author of this book is well aware about the negative attitude of some scientists of modern physics for the existence of the substance Aether, still It is explained in a relevant chapters that on the ground of reality there is no reason to deny for existence of Aether. This type of Brahm is neither imagined nor denied. Actually, Brahm theory itself is not just a substance Aether theory, instead it is a theory of dynamic flux of momentum carrying units forming push gravity with Aether as a manifestation. Here substance Aether is not a thing existing just between masses or just penetrating masses, but it actually forms masses within it. Masses are nothing but complex waves within

Aether only. This Brahm including Aether may replace a space time matrix to some extent without supporting calculations of special relativity.

In the upcoming Chapters, I have described why the possibility of existence of Brahm is must there, and what's wrong in the rejection of the existence of Aether based on the unproved theories and some misunderstood observations. So, it is a priority to understand aspects which are considered as not in favor of the existence of Aether. On this way, the book proceeds with fair introduction of the "Brahm-Theory of the Nature" on a relatively logical and different ground.

No single theory of modern science is capable of explaining most of the things all together. We can say that none of the existing theory could be a complete truth. Author didn't try to stay away from any logical contradictions, instead try is made to bring all doubts together and provide the possible explanations. Author is not interested to support any theory that don't provide a good sense to understand the Nature properly. Separate isolated restricted approach which is common in modern physics is not entertained here, Instead the theory describes here a whole concept that must provide a common platform for each and every existing thing. Still this theory provides options where exact predictions remain obscure, like choosing between state of Aether as liquid or solid, choosing between a nature of light as a wave or mass. There may a priority be given to one of such options, but deterministic approach is not carried out to keep acceptance for any obscure possibility. There are many predictions of this theory different from the prediction of the modern science. The theory predicts different possibilities for the structures like core of the big planets, core and life cycle of stars, nature of big heavenly bodies, structure and function of galaxies, quasars, and nature of almost every existence, different view for the centre of the galaxy. Ultimately this theory helps to calibrates a common sense in a manner that makes it clearer, logical and realistic. This theory contradicts with the calculations of special theory of relativity, still this theory of Brahm explains observer specific constancy of speed of events, time dilation and maximum limit of speed for mass. This theory of Brahm don't agree with relativistic singularity, relativistic blackhole based on relativistic vicious gravity. Shadow gravity model doesn't allow blackhole formation instead it demands less gravity in the centre of Galaxy. This theory accepts Aether and Brahm everywhere and here there is an absolute reference frame of Aether unlike no absolute reference frame for motion in special relativity. Brahm is a dynamic reference frame that provides something like a relativistic neo-aether which is there in a space that defines time and space curvatures to form gravity. Brahm or Aether as a universal reference frame is accepted here, that is against the postulate of special relativity. So, Relativity with all its calculations and predictions is no more a theory that could co-exist with this framework but few modifications in relation to Brahm may survive relativity within the framework of this Brahm theory. This model also rejects the idea of extra spatial dimensions other than 3 spatial dimensions, relativistic blackhole, wormhole, graviton, gluon, string theory, and many concepts of quantum mechanics. This model also disagrees with conventional concepts about charge and electricity. This theory can't provide the insight for choosing between an expanding and non-expanding universe. Both could be possible within this framework. But a view that supports nonexpanding steady state universe is described in a book. If we want an expanding nature of the universe then we can assume pulsatile fluctuations within aether at a huge volume that can accommodate the whole universe with reciprocal fluctuations in surrounding volume.

This Brahm theory allows more fundamental questions which are not entertained much in a modern science. In Routine, we feel satisfied by giving name to observation without properly understanding a cause behind the phenomenon. This theory tries to trace the cause to the deeper level, that can explain every phenomenon as a mechanical concept. This theory rejects the possibilities of existence of any force like pure attraction at all. The only way remains is to explain gravity and all other forces as pure mechanical interaction that must obey laws of classical mechanics in a physics. Another example of such widely ignored question is, "what is inertia?" We know the law of inertia, but still inertia is an unsolved mystery. So, phenomena are questioned beyond the conventional level, and an attention is drawn to the existence of poorly expressed things.

First part of the book includes the history of Aether (existence of something in a so thought blank space that affects masses), its rejection in a modern science and possibility of its existence and supports for existence of Aether in the modern physics.

Second part covers a proper Brahm theory, that describes core concept of the theory. In second part explanation of the various fundamental phenomena are there, which include the formation of matter, gravitational and other forces. Nature of light, nature of matter, nature of galaxies and big celestial bodies are also described. ***The description and explanation of the material world with cause behind it is there in a second part.*** There is some gap when trying to progress from fundamental mass to the atom level, and that is an unexplored and poorly understood portion of this theory and that is one of the reasons why this theory doesn't answer some questions and mysteries. But it doesn't mean, this theory can't explain subatomic particles. This just means that calculation for exact anatomy of quantum object is not provided. This suggests further development of the theory by including data of experiments.

The Special theory of Relativity needs modification with this framework, so not all predictions of special relativity are accepted within this theory. A chapter 19 that shows relativity in relation to Brahm is there in a second part of the book that provides an insight for understanding relativity in relation to Brahm.

This theory creates a mechanical sense. Here mechanics also includes fluid and wave dynamics along with motion mechanics. This book strongly states that *"A mechanical explanation of any existence or phenomenon is the only fundamental real explanation of that existence or a phenomenon"*.

Dr. Hasmukh Devidas Rathod.
Date: 18/5/25

Part I

History of Aether, and its importance

BRAHM

1. The history of Aether

Before describing the history of Aether I want to clear some points. Is this book about Historical Aether? Actually not. This book describes "Brahm" which can be identified as Aether with its dynamics, but it is not just assumption like a historical aether in science and philosophy. It is more than just Aether. Actually, Aether exists within "Brahm". One clear point is that it is the Natural existence (Nature) that harbours everything including infinite number of universes. It is prevalent everywhere including at so-called blank space also. Historical Aether was the concept of an existed substance in a so-called blank space between masses. This was against common belief of Newtonian physics related to blank space. This is the relevance between historical aether and Brahm described in this book. That's why history of an Aether is described here as a view in contrast with absolute Newtonian blank space and no-aether concept of the special relativity.

Before seeing the history of luminiferous Aether in science, let me show you some history of "Brahm" a supreme existence or the omni existence in Indian philosophy.
Let me show you the history of Brahm or the medium Aether on religious and philosophical grounds first.

A brief history of Aether in philosophy and Hindu religious texts.

Brahma-tattva (Brahm or Brahman) (A supreme element or supreme being)

The word Brahm or Brahman is taken from the philosophy of the supreme aether like existence that harbours each and every existence within it as a physical world which is its manifested form powered by "Maya". Brahman is a singular form of word which represents a single entity and description of this Brahman as an individual with will power and consciousness is there in Veds. Here I have used the word Brahm instead of Brahman to describe that existence as a physical existence and not as a conscious being. This description of Brahm is there in a Hindu ancient philosophy known as Advait Vedanta means a philosophy of non-dualism. Here Theory of Brahm don't include the spiritual aspect of the Advait Vedanta. There are disputes about Brahm as material form or a conscious being. Soe texts in Hindu literature favours material form only while other includes Brahm as conscious being. I would like to describe both as a history of Aether or Brahm.

I shouldn't deny that religious texts have their context about origin of the universe. In my childhood the only source of knowledge to satisfy my curiosity was my father.

He introduced the concept of "Ishwar", the supreme existence as follows.

"It doesn't have general physical properties like that of material things of the world, including colour, smell, visibility, taste, texture etc. Such properties can't be applied to the "Ishwar" in its Neerakar (Unmanifested) form, thus, Ishwar is also called Neergun (without physical property like that of material things). Ishwar doesn't get affected by wear and tear. It can't be burnt-out; it can't get destroyed anyhow. It is everywhere and everything is manifested within it. It is in the living and in the non-living things. Every existence is its "Maya" (it's manifestation). Gods we worship are not Ishwar but a manifested form of Ishwar, Neerakar Ishwar is beyond All these Existences. I questioned If we can't see or feel it's existence then how we may know about it! Father replied, Gyani (Knowledgeful) Peoples said that it can be realized with wisdom and knowledge. It can't be seen with Physical eyes, but it can be perceived with eyes of knowledge "Gyaan Chakshu". So, this was the first introduction of Aether as "Ishwar" to me by my father, but he told me all these things as a matter of faith. I didn't know about the source of his faith and knowledge, but later on I found that the content is same as described in the Nasadiya Sukta of Rigved, Advait Vedanta and Advaita philosophy of Adi Shankaracharya. Maharshi Dayanand Sarasvati and Swami Vivekananda popularized these philosophies in modern era.

In Hindu Sanskrit literature Rugved (Rigved), following descriptions are given.

The Nāsadīya Sūkta also known as the Hymn of Creation, is the 129th hymn of the 10th mandala of the Rugveda. It is concerned with cosmology and the origin of the universe. It begins rather interestingly, with the statement – "Then, there was neither existence nor non-existence."

English translation of Nasadiya Sukta (Hymn of non-Eternity, origin of universe):

There was neither non-existence nor existence then;
Neither the realm of space, nor the sky which is beyond;
What stirred? Where? In whose protection?

There was neither death nor immortality then;
No distinguishing sign of night nor of day;
That One breathed, windless, by its own impulse;
Other than that there was nothing beyond.

Darkness there was at first, by darkness hidden;
Without distinctive marks, this all was water;
That which, becoming, by the void was covered;
That One by force of heat came into being;

Who really knows? Who will here proclaim it?
Whence was it produced? Whence is this creation?
Gods came afterwards, with the creation of this universe.
Who then knows whence it has arisen?

Whether God's will create it, or whether He was mute;
Perhaps it formed itself, or perhaps it did not;
The Supreme Brahman of the world, all pervasive and all knowing
He indeed knows, if not, no one knows

—*Rigveda 10.129 (Abridged, Tr: Kramer / Christian)*[3

(source- Wikipedia)

So, in a Nasadiya Sukta there is a description of the existence before the creation of a physical world.

This describes something which have a state between being and not being. There should be a question in a mind that how something could be an existence between being and non-being! I like to give an example to explain this. Suppose there is a TV screen. If it is on and it shows a picture, and we recognise that picture as something. If TV is off, then we say that there is nothing on TV screen. Here TV screen is there still I shows nothing on it. Another example is an ocean with waves. If we describe waves as something, then ocean without waves carries nothing. Similarly motion within Aether forms physical world which contains masses and waves. So, here physical world is the manifestation within Aether which is called physical existence or Srushti. Aether itself in relation to physical world is not a physical existence because it remains unmanifested as a physical thing. We can imagine Aether as a one omni existence without any discrimination in it when it doesn't form a physical world within it. According to Nasadiya Sukta Heat created an event and that resulted in the manifestation of the physical world. We can imagine this like a wave formation in a medium due to motion of units that forms discriminated things within medium and it forms the physical existence. In Advaita Vedanta it is described that physical world is a manifestation of Brahm mediated by Maya.

This Sukta described above states that there was an omni existed thing which has a no discrimination within itself. This is termed as Xir in Vishnupuran. Also called "Paramtattv" (Supreme Element). Physical world is a manifestation within it. It is also described as a fluid or water. This Sukta states that before creation of the world that Xir was lonely and alone, everything was created then after, even Angels and Gods were created then after, So, nobody knew how all that happened and came to existence, neither Gods nor Angels knew it. We could only imagine possibilities. It may be the willpower of the Ishwar that led to the creation of the physical world or something else, we can't know.

But I heard that as a description in the Advait Vedanta that the Ishwar creates physical world within own self and then dwell in it to experience it, which I found as a part of scriptural debate between Ashtavakr and Acharya Bandi in the "Mahabharat" as a famous episode in the Vana Parva (Book of Forest). Let me represent this debate.

Acharya Bandi was highly learned and arrogant court scholar of King Janaka. He had reputation for defeating all challengers in philosophical debates and drowning them in the river (Actions were reverted at the end). Ashtavakra a young sage, physically deformed in eight places (hence his name "Eight bent). Despite his appearance, he was exceptionally wise and spiritually advanced. His father was also a renowned scholar.

Let me provide that part of debate as a question answer between those two. Ashtavakr as "A" and Acharya Bandi as "B". There are various forms of this debate, but I am representing a popular one with description of Brahm in it.

A: Where has the material world come from and how?

B: This world has come from him and him alone. Who is he?

A: He is the "Brahm (Brahman)" the ultimate being, He is the "Ishwar" who forms this world with his power called "Maya". He alone is the creator, the sustainer and the destroyer of this world.

B: How does Brahm create, sustain and destroy this world?

A: the way a spider makes its web, moves about in it and later swallows it, in the same way Ishwar creates, sustain and destroys the world.

B: What is a living thing?

A: Living thing is a Soul (Atma) and although it is changeless, under the influence of misconceptions, it mistakenly starts considering itself to be the mind and the body. That's why it can experience this world.

What is the misconception?

B: Considering non self to be the self, considering the changeless as changeable and considering this world to be real is the misconception.

Then what is the Realization?

A: The knowing of self is Realization, it is the Realization that free the one from conflicts, illusions and fantasies. That free us from feeling duality, idea of separate being from others. Realization gives an insight that enable the human to recognize himself as a Brahm.

What is the way or procedure to know Brahm?

B: There are two ways to know Brahm. First one is knowing what Brahm is not and excluding that. After excluding names, forms, qualities, flaws, changeable things knowing what is not changeable. Second way is recognizing the truth as it is. It is the nature of the existence. The world has no existence without it. That was Brahm, that is Brahm, that's why all are Brahm. We all are nothing but a Brahm. The whole world is a Brahm.

A: What is the procedure to know Brahm?

B: It can be known by appropriate social behaviours, spiritual thinking, listening with attention, rethinking on daily experiences, analysing conclusions and by entering in a special state of consciousness known as Samadhi.

What are the traits of person knowing Brahm?

A: If anyone claims that he completely knows Brahm then actually he has not understood Brahm. As soon as a person knows Brahm, the conceit of knowing Brahm vanishes.

B: Can it be understood with Logic?

A: No, but logic can be helpful.

Can it be understood with prayers and devotion?

B: No but prayers and devotion can help.

Can Brahm be known through yoga and thinking?

A: no, but yoga and thinking can be helpful.

Debate continues for a long time. The climax of the debate comes when Ashtavakra poses the riddle: "What is that which is one, but acts like two? What is that which is two, but is essentially one? What is that which is neither two nor one? Acharya Bandi could not answer this question and defeated.

[this debate is taken from the you tube video with the link https://youtu.be/IR9QwP4boiE?si=H5xFHQ_ySoZynEmO

Upanishad Ganga 1 Ep 04]

This scriptural debate is about ultimate existence and supreme being. The debate underscores the Advaita Vedanta philosophy of non-duality, emphasizing the ultimate oneness of reality. This debate refers to the concept of one reality and effect of Maya that creates illusion of difference.

If we see it as an origin of material world with natural processes without interference of active conscious interference, then it matches with the "Aether-Brahm" theory described in this book. We can note that there is a no confident view of role of active will power in a creation of the world in Nasadiya Sukta of Rugved, So, we may consider this debate as a personal view of the writer of the Epic Mahabharat with belief in an ultimate conscious reality, Brahm. To avoid spirituality the theory in this book neither involve consciousness of Brahm nor this theory of Brahn rejects any characteristic of Brahm beyond the scope of this theory.

In a Vishnupuran, description of waterlike Substance prevalent everywhere, is termed as "Xir", which is a liquid like a water, then there was a spontaneous event as rise of vibration within it, which spreads everywhere and what created is the physical world.

"Advaita-vâd" a believing in one supreme entity, which is described as the "Brahm" in Veds.

Its description in Hindu literature describes following qualities.

Few words represent these qualities.

Sada, Sarvada, Sthayi, Satya, Shashvat, Samarth, Sampoorn, Sarvagya, Sarv-vyapi, Sarvasva.

Sada, this word means Ever. It is the existence since ever and it will remain forever. Sarvada means a thing forever. There is a no birth or creation of it and there will not be the destruction possible. Aja is the world used for Brahm which means, it is unborn. Amar is the word used for a quality Brahm which means an existence without a possibility of death.

It is "changeless" (Nirvikâra or Ajara). Primal cause is changeless, what we found as a change is just its state. There is a creation and demise of the physical world within this changeless primal cause. The changeless quality is also described with the word "Sthayi"

"Omnipotent" with all power (Samarth): each and everything existed is a manifestation within that all powerful primal presence.

"Omnipresent" (Sarv-vyapi) Brahm is all pervading and present everywhere because everything is a manifestation within it, so-called "Maya" which is not an illusion but a real thing still it is a manifestation of absolute real existent "Brahman" and not thing itself.

Omniscient: It is also considered as "all knowing" (Sarvagya)

Everything is within it including Maya, it is the only existence (Sarvasva) or omniexistance.

So, it is

Sada, Sarvada	Forever, always existing
Aja,	Unborn
Sthayi or Ajara,	without wear and tear
Amara,	Immortal
Samarth,	Omnipotent
Sarva-vyapi,	Omnipresent
Sarvagya,	Omniscient
Sarvasva.	Omniexistance.

This is the description of that root cause in the Vedik literature of Hindu religion. Description of Brahm in this book is little bit more realistic, that doesn't include some of these supreme qualities or power. So Brahm described in a "theory of

Brahm" is not omnipotent or all powerful, it is the natural existence medium that allows every physical exitances to exist within itself as manifestations within it due to all natural processes which are just consequences and not an intended work. Consciousness of this omni-existence is a character described in a Veds but not considered as a necessary character of Brahm described in this book, for to be manifested as a physical world.

So brahman in Veds is in a state between alive and dead and also conscious. But a Brahm described in the theory of Brahm in this book is just composed of two things, something (mass) and nothing (blank space) which are manifestations in Aether a matrix or medium.

Consciousness of Brahm is not included in this theory, but natural mechanisms are described to understand a self-maintaining system without need of any consciousness. According to this theory of Brahm, Life is a quality in the living thing which is a creature developed in physical world which may carry a system known as a nervous system and consciousness is a real-time partial awareness of self and surrounding which is manifested within nervous system only powered by electrical activity of the nervous cells, neurons. This books only describes the physical world without life and don't include philosophy related to life or consciousness in this book.

So, this book doesn't describe a Brahm as a conscious thing, but consciousness is one of its manifestations which is a character of living things. Still this book doesn't rule out conscious things at other levels and don't say anything about omnipresent omnipotent Ishwar.

Sanskrit shloka from Rigved about "Ishwar"

ईशावास्यमिदं सर्वं (Isâ-vâsya-mîdam-sarvam)

Subhashita (well spoken)

ईशावास्यमिदं सर्वं यत्किञ्च जगत्यां जगत् ।
तेन त्यक्तेन भुञ्जीथा मा गृधः कस्यस्विद्धनम् ।।

English transliteration

īśāvāsyamidaṃ sarvaṃ yatkiñca jagatyāṃ jagat ǀ
tena tyaktena bhuñjīthā mā gṛdhaḥ kasyasviddhanam ǀǀ

English meaning

This universe, made of living and non-living things, is permeated by Ishwar. Humans may use the resources as needed but may not hoard as if 'this all belongs to me'.

Other meaning can be also there where we can define Isa as a motion means an energy. This would turn the meaning as follows.

Motion is present at everything; that's why the world is twinkling. By losing that (motion) annihilation occurs and leaves no place for the existence

Source
Ishopnishad 1.

This shlock defined Ishwar as a thing that permeates every living and non-living things.

From folk culture in India.

Four lines of "Vishvambhari stuti" (song of the worship for the Goddess Durga) states.

"ખાલી ન કોઈ સ્થળ છે વિણ આપ ધારો,
બ્રહ્માંડ માં અણુ-અણુ મહીં વાસ તારો,
શક્તિ ન માપ ગણવા અગણિત માપો,
મામ્પાહી ૐ ભગવતી ભવ દુ:ખ કાપો."

Khâli na koi sthal chhe vin aap dhâro,
Brahmând mâ aṅu aṅu mahi vâs târo
Shakti na mâp ganavâ aganit mâpo
Mâmpâhi Om Bhagavati bhav dukh kâpo

Translation: "There is a no empty place without you is a rule, you reside in each and every atom of the universe. Plenty of units are there for measurement, but the energy is beyond measuring limits. Hail Goddess, please deduce our sorrow due to life."

So, this is the commonly sung worship song in one of the states of the huge country India describing its omnipresence and stating that nothing is empty, I can't find details about writer and time of origin, but it may be few centuries old.

So, most of the texts including Veds, Upanishads, Epics like Mahabharat and Ramayan, describing Brahm is the oldest in the world and it has description of supreme existence everywhere, responsible for existence of each and everything and describing that everything is just a manifestation within that supreme existence.

I haven't read original source of these texts in detail as I don't know Sanskrit, But I came to know about this description mainly after internet became widely available in India. I have heard somewhere a story about tortoises supporting earth in Christianity. I have seen many videos in Hindi language describing the Vedic Science theories. So, description of omnipresent supreme existence is ancient.

So, these were the descriptions in ancient texts, which have relevance with the property of Brahm-Aether.

There can be much more or too much in Sanskrit literature, but I don't have access to it as it is in Vedic Sanskrit which is still not properly decrypted. Acharya Agnivrat Naistik have translated a Brahman Granth "Aetrey Brahman" with his own insights that resulted in a Vedic Physics book known as "Ved Vigyan Alok" which represents a Vedik Rashmi theory. I haven't included that philosophy in this book because it includes conscious interference and command of Supreme conscious being. That describes "Rashmis" as a governing force behind all forces and Manifestations. There is a complex description of the mechanism described in his books which is not a mechanics at least. I have superficially read some chapters to know the essence of those E-books which are translations of the Aetrey Brahman Granth.

The word Aether is from the Greek language, and it means a Fresh Air of clear Sky. It is also personified as a deity, Aether, the son of Erebus and Nyx in traditional Greek mythology.

But the expression in this book is more similar to "Xir" or "Brahm" in ancient Indian literature, that's why this book has the title "Brahm". I used the word Aether as synonyms for "Brahm" because it is common similar English word being used in physics, but it is Brahm and not a thin luminiferous Aether imagined for the propagation of light.

Influence of Advait Vedanta by Swami Vivekanand on Nikola Tesla:

Swami Vivekanand was a reputed young Indian saint of Sanatan Hindu Dharma, who is Best known for his groundbreaking speech in the 1893 at World's Parliament of religions on the site of present-day Art institute, in which he introduced Hinduism to America and called for religious tolerance and an end to fanaticism. Swami Vivekanand gave his first lecture on 11 September 1893. Swami Vivekanand sailed from Madras to Chicago to participate in that event. He introduced Vedanta and Yoga to Western world. Theory of Advait Vedanta about Brahm as a supreme entity and the world as its manifestation as a "Maya", impressed Nikola Tesla. Mr. Nikola tesla used ancient Sanskrit terminology in his description of natural phenomena. Nikola Tesla described the universe as a kinetic system filled with energy which could be harnessed at any location. During following years his concept were influenced by Swami Vivekanand's teachings. Nikola tesla believed that the space is not blank, but it is a full of mechanical energy even before he became familiar with Vaidik Philosophy. After meeting Swami Vivekananda, Tesla was influenced by eastern Advaita Vad. He started using words like PRANA and AKASHA to describe the concept of Luminiferous Aether as a source of formation of matter and carrier of energy. Nikola Tesla also shared his ideas with Kelvin.

Swami Vivekanand expected mathematical solution for Advait Siddhanta that "force and matter are reducible to potential energy" from Nikola Tesla. Swami Vivekanand was hopeful that Tesla would be able to show that what we call matter is simply potential energy because that would reconcile the teachings of the Veds with modern science. Tesla failed to provide such solution, and a dream of Swami Vivekananda was not fulfilled.

My Opinion on eastern (Indian) Philosophy.

Ancient Sanskrit Language is still not decoded or decrypted. Today's Sanskrit is a new and different language from ancient (Vedik) Sanskrit. The source of knowledge, the ancient Sanskrit is a lost language. Even translations of known mantras are also incorrect with different views of Acharyas. The detailed theory of Brahm or Brahman a supreme spirit is lost, and only remnants are there in Veds and Vedants as philosophies. Ancient Rishis tried their best to harvest the real knowledge from the ancient old Sanskrit, but in the absence of technology and advanced science they failed to recover all. This leads to many misunderstanding in Indian religion and mythological beliefs. After understanding the theory of "Brahm" described in this book, may provide some insights to recollect the massage hidden in Indian mythology. The word "Brahman" is more common than Brahm in Vedic literature, This Brahman is the Ishwar who is a conscious, omnipotent, omniscience and omniexistance. That's why I used term Brahm instead of Brahman which don't represent an individual like single solitary existence, but it represents a substance in a state instead. Before the method of documentation as a written text there was a long oral tradition of conveying knowledge as mantras and hymns. When the method of documentation established, many Acharyas tried the documentation and translations (Bhasya). What was clear as a concept like omni existence, all pervading, all prevalent were grasped and translated with accuracy but complex things were misunderstood mostly and that's why Advait Vedanta remained as a philosophy instead of a scientific theory. The detailed incorporation of phenomenal facts in Hindu Mythological characters and paintings (generated from a written descriptions in Veds and Parans) points out to a much more developed Brahm theory in addition to philosophy of Advait Vedanta. Existence of such developed scientific theory in ancient time is a mystery.

Brahman

The Possibility of Brahm as a Brahman, an existence of Brahm as an individual being with consciousness and will power, is neither supported nor denied in this book. But this book provides insight to the factors not in a range of a modern physics that may support the existence of such powerful Ishwar. So, Ishwar is term that defines an Aether and Brahm with consciousness and willpower.

Some supreme characters like Shiva (primal mass), Natrâja (spinning torus), Devi Shakti (kinetic energy), Krishna (centre of galaxy) and some events like Samudra-Manthan (Dynamo), Kamdev with bow and arrow (cathode rays), Devi Chhinn masta (principle of closed circuit), Ganga avtaran (manifestation of cosmic stream of energy) etc. are described as personifications of physical real exitances and scientific phenomena. This are described in brief in a chapter "Relevance with Indian mythology"

A brief history of Aether in the mainstream science at a glance:

Mainstream science was philosophy before few centuries. Still a Modern fundamental physics is full of imaginations and philosophies. Plato stated that "there is the most translucent kind which is called by the name Aether". Though he had adopted the classical system of four elements. Aristotal a student of Plato at the academy, stated that there are celestial spheres which are made up of Aether. Those are responsible for circular orbits of the celestial bodies like stars, sun, planets. Plato considered it as the first element but later on it was considered as fifth element and also called it Aether, a word that Aristotle had used in "on the heavens and the meteorology" but that concept of an Aether was restricted to its function related to motion of celestial bodies like stars and planets. That Aether was considered different from four terrestrial elements. That Aether was assumed to have a natural motion in circles only and made a celestial sphere that held stars and planets. That Aether filling gaps between planets was believed less dense than a bodies of the planets.

In the medieval era the acceptance of Aether was there. Aether was also known as the fifth element or quintessence; it was believed as the material that filled the space beyond the terrestrial sphere. But that was not thought as an origin of the mass. This philosophy developed independently different from Indian philosophy within western countries because Asian languages were not readable for them. Still, it seems like a "Charvak Darshan" based on four fundamental elements and rejecting fifth element earlier and later on returning to "Panch bhut" a five-element philosophy in Hindu philosophy (Veds and Upanishads), Ayurveda and Yoga. In contrast to Vaidik belief of fluidlike substance, their imagination about Aether was of very less dense substance as an Aether that fills the blank of the space. Robert fludd (1574-1637) an English Physician and alchemists stated about the density of Aether, and he said that Aether is "subtler than light". He believed it as a penetrative for matter, and it-self Aether is a non-material thing. Though other ideas of fludd were not correct, like his belief in geocentric world and believing that earth is the heaviest body in the world. So, he didn't carry a good top view of the world. We can't consider these philosophies as a progress because those were inferior to ancient philosophies, but we can consider it as a good comeback after a fall from the height of philosophical knowledge.

Though it is commonly believed that Newtonian physics accepted space as a blank void, but Newton also imagined Aether as a medium responsible for light propagation, cause for gravity and responsible for behaviour of light in optics. Meanwhile *Nicolas Fatio de Duillier* a mathematician and a friend of Isaac newton developed a *Shadow gravity theory*, which provided the mechanical explanation behind the gravitational force. According to me that was *"the great explanation ever given for the mysterious force called gravity"*. It was the pure key of the fundamental force governing the whole aethereal universe, grasped by someone in known human history first time. In that theory he imagined very small invisible particles in a space called inframundane particles, that move in every direction in the space continuously and that strikes with the material bodies and create an effect which is perceivable as a gravity. Thus, Fatio found the relation between gravity and something in the blank space which is moving and carrying energy. This theory is also known as "the kinetic theory of gravity". He stated directly that the blank space is not actually blank. He imagined a blank space as a gravitational environment, indirectly in a different manner he represented not Brahm itself but one of the properties of Brahm described in this book. His idea was ahead of his time and even ahead of the present modern time. That did not gain any attention of mainstream science.

Johann Bernoulli II, a Swiss mathematician, proposed a theory in 1752 that Aether was a fluid with many tiny whirlpools that permeated all space, and these whirlpools give elasticity to Aether. Bernoulli's theory was popular during 17th and 18th centuries as a way to describe the motion of light. According to this theory Aether was a fluid with many tiny whirlpools that allowed Aether to transmit vibrations from light corpuscles as they travelled through it. This theory has a close similarity with some part of this Brahm theory.

Again, kinetic theory of the push gravity came in a highlight when Lesage tried in vain to revive and reintroduce that same theory. Kelvin published a note on Lesage's model in 1873, in which he found Lesage's proposal was not compatible thermodynamically and he also suggested to improve that model using then popular "the vortex theory of atom".

Kelvin concluded about Lesage's kinetic theory,

"This kinetic theory of matter is a dream, and can be nothing else, until it can explain chemical affinity, electricity, magnetism, gravitation and the inertia of masses of vortices. Le sage's theory might give an explanation to inertia of masses, on the vortex theory, were it not for the vortex theory, were it not for the essential aeolotropy of crystals, and the seemingly perfect isotropy of gravity. No finger post pointing towards a way that can possibly lead to a surmounting of its flank, has been discovered, or imagined as discoverable."

I think they were nearer to "the theory of Brahm" described in this book, but Michaelson Morely experiment, and special theory of relativity have changed the path of the progressing science on the way of progress of course to collect the concept of time dilation.

"The vortex theory of atom" was a theory proposed by William Thomson better known as Lord Kelvin. In that model he imagined atom as a vortex in the fluid medium Aether. So, Aether model for explaining mass as nothing but a vortex in the liquid medium is not new. It is already tried approach, but complexity of the reality ruined the simple form of that theory. In that theory atom was imagined as a primary vortex loop in a fluid Aether at atomic level, which did not stay competent with experimental results and after discovery of the electron and other subatomic particles, that theory was made abandoned.

Larmour and Lorentz electron Aether theories are well-described Aether theories. But primary unit of Aether in those theory was assumed as an electron, which was not explained in further detail. These theories were failed when quantum physics and relativity contradicted with them.

Few lines by Lamour.

The Science world is increasingly dominated by scientists in favor of the Bohr Heisenberg interpretation of physical reality. Physical theories still required visual representations and illustrations to be completed.

Sir Joseph Larmor was the first person to calculate the rate at which energy is radiated by an accelerated electron, and the first to explain the splitting of the spectrum lines by a magnetic field. His theories were based on the belief that matter consists entirely of electric particles moving in Aether.

The acceptance of Aether as a luminiferous substance become prominent in the 19th century, when it is accepted that light has a wave nature. That was believed that There must be a medium for the wave to travel along with. So, in the 19th century mainstream science has acceptance for the existence of luminiferous Aether which is responsible for the propagation of the light. It was prime priority of physics in the 19th century to find out property and nature of the luminiferous Aether.

Three possibilities of Aether depending on the drag of Aether along with moving masses was plotted as Dragged Aether, partially dragged Aether and Stationary Aether.

1. Dragged Aether: Mass drags Aether within its vicinity. This vicinity is the area around the mass up to that extent Aether is being dragged hypothetically.

2. Partially dragged Aether: Though mass drags Aether but not completely, only partial drag is there within the vicinity of the mass. There is a relative motion.

3. Stationary Aether: Mass don't drag Aether but moves relative to that Aether in the space during its motion in the space, thus this stationary Aether forms a one single absolute reference frame for motions relative to it.

There was an observation called "aberration of Star light". In this phenomenon stars appeared to be displaced from their position according to earth's position and motion in the orbit around the sun. This happens due to the relative motion of the earth to the beam of light. This phenomenon was in support of stationary Aether and not in a support of dragged Aether with earth vicinity. Because light was assumed as a wave in Aether, and it was argued that if Aether being dragged with the earth, then the phenomenon of aberration of the star light should not be there due to drag of light along with Aether in a vicinity of the earth.

After observation of this phenomenon the only possible possibility for the existence of Aether remained was a possibility of a stationary Aether which don't get dragged with the earth and remain stationary and thus moving in relation to earth.

During that period **Albert Michaelson** invented the apparatus called the interferometer. He did some experiments with that apparatus to find out Aether wind causing a drag on the light wave. He performed alone some experiment then collaborate with **Edward Morley** and performed Most famous failed experiment called **Michaelson-Morley experiment**. Conclusion of this experiment was that there is no Aether drag on the light, means there is no existence of the stationary Aether. Together with the conclusions from the observation of the aberration of the star light this experiment provided the conclusion that there could not be the existence of luminiferous Aether either stationary or dragged with the earth, capable of producing drag effect while relative motions crossing the path of light. This experiment also provided the conclusion that speed of light don't get affected by relative motion of any Aether like medium and remains finite and constant for the observer. This Conclusions became the base for the special theory of relativity. Hendric Lorentz tried to explain the result of Michaelson Morley experiment with assumption of Aether as an electron field made up of electrons. But he failed to explain it. Lorentz might be inspired from Voigt's transformation equations about doppler effect of waves in a medium and Lorentz developed Lorentz transformation equations in 1904, that gave the concept of time dilation and length contraction. But due to the assumption of an Aether as an electron field he was unable to explain the behaviour of light and relativity properly because there was a universal reference frame made up of electron field. In a 1905 A scientist impressed by the work of Lorentz used Lorentz transformation, Time dilation, Length contraction concepts, but denied existence of Aether in any form. He built on Lorentz's ideas, including Lorentz's transformation, which described how differently observers see the same event. He also disagreed with Lorentz's theory of an ether. He provided the science world with a postulate of observer specific constancy of speed of light and concept of the inertial reference frames. He defined time as the fourth dimension and described space as a 4D space time Matrix. He explained all these things in a theory called the Special theory of relativity. The name of this physicist was Professor Albert Einstein. He also used Minkowski space concept in his theory later on.

Though the interchangeability of relatively moving subjects creates a logical and mathematical errors on the ground of symmetry and twin paradox completely ruins out the theory, mainstream science still managed to believe in the special theory of relativity and considered it as a ground for the rejection of the existence of the luminiferous Aether. In 1915 Professor Albert Einstein further developed the theory and included the effect of an acceleration and gravitational field on the light path and provided a concept of geodesics and curvature in a space-time matrix by mass. This theory is called General Theory of Relativity. This theory stated that gravity is due to curvature of space-time matrix around the mass. This curvature is formed by mass with ability which is an internal property of the mass. Then Lorentz wrote the letter to Albert Einstein asking, did he reintroduced Aether as a space-time matrix, which he had rejected in a special theory of relativity? Albert Einstein gave the answer with the explanation that this is a Neo-either, no "substance" and no state of motion can be attributed to that new Aether; space time matrix is just a mathematical concept to derive gravitational field. That was a poor answer because it was simple to understood that if something is being curved than it can't be just a mathematical tool but must be something real that may be a dynamic presence at least, if not a steady or moving substance.

Physicist Robert B. Laughlin wrote:

It is ironic that Einstein's most creative work, the General theory of Relativity, should boil down to conceptualizing space as a medium when in his original premise [in special relativity] was that no such medium existed. The word 'ether' has extremely negative connotations in theoretical physics because of its past association with opposition to relativity. This is unfortunate because, stripped of these connotations, it rather nicely captures the way most physicists actually think about the vacuum. Relativity actually says nothing about the existence or nonexistence of matter pervading the universe, only that any such matter must have relativistic symmetry. It turns out that such a matter exists. About the time relativity was becoming accepted, studies of radioactivity began showing that the empty vacuum of space had spectroscopic structure similar to that of ordinary quantum solids and fluids. Subsequent studies with large particle accelerators have now led us to understand that space is more like a piece of window glass than ideal Newtonian emptiness. It is filled with 'stuff' that is normally transparent but can be made visible by hitting it sufficiently hard to knock out a part. The modern concept of the vacuum of space, confirmed every day by experiment, is a relativistic ether. But we do not call it this because it is not accepted (taboo).

Louis de Broglie stated, "Any particle, ever isolated, has to be imagined as in continuous "energetic contact" with a hidden medium."

However, as de Broglie pointed out, this medium "could not serve as a universal reference medium, as this would be contrary to relativity theory."

So, we can see here that, Relativity theory is the obstacles that don't allow scientists to accept the perceivable quantum vacuum as a universal reference medium aether. This is because a space as a substance in a vacuum can't have a relativistic symmetry due to its stand as a universal reference frame.

We will see ahead in part two of this book that Brahm is dynamic environment with Aether medium, and it have symmetry that can explain time dilation also, still it contrasts with the special relativity and we will not try to rescue relativity instead we will try to understand Brahm and its consequences independently.

Quantum vacuum:

Quantum mechanics can be used to describe space as being non-empty at extremely small scales, fluctuating and generating particle pairs that appear and disappear incredibly quickly. It has been suggested by some such as Paul Dirac that this quantum vacuum may be the equivalent in modern physics of a particulate Aether. However, Dirac's Aether hypothesis was motivated by his dissatisfaction with quantum electrodynamics, and it never gained support from the mainstream scientific community.

So, we can see that,

The following points are in support of Brahm-Aether.

1. Observations of quantum physics.

2. Space matrix, geodesics, curved space concept in a General Relativity, a (neo)-ether concept, time dilation.

The following are against Aether.

1. Special theory relativity.
2. Combined conclusion from, aberration of star light phenomena & Conclusion derived from the Michelson-Morley experiment.

So, before the introduction of Aether again, one has to pass through this two major obstacles.

In this book Aether is described as a medium that forms everything within it as waves and complex vortices of waves. It is not a substance only between two masses. It doesn't contradict any observation or theory, and it fulfills the gap. Brahm is described as an existence carrying such medium in this book. That Aether is not a substance just between material objects, but it is a medium that carries mass within it as a complex compound vortices of aethereal waves. So, during motion of such mass, mass rides on aether as a wave complex. Thus, no room for drag or friction remains there. There is an interaction of matter with gravity forming flux, and that regulates speed of matter in Brahm and provides inertia.

Some parts of this chapter are taken from the following references.

Reff.
https://en.wikipedia.org/wiki/Aether_(classical_element)
https://en.wikipedia.org/wiki/Aether_theories

2. Rejection for the existence of the substance Aether in a special theory of relativity

As, we have seen in the chapter 1 that, there are three main reasons behind the rejection of existence of luminiferous substance Aether.

1. Conclusions from the experiment of Ole Rømer and observation of the binary star system.

2. Conclusions from the aberration of star light phenomena

3. Conclusion derived from the Michelson-Morley experiment.

The luminiferous Aether was imagined as the medium for the propagation of the light as a wave. The behaviour of the light wave during its travel in the space is observed and that provides some information about the properties of the light. One of them is the finiteness of speed of light in a vacuum. Many scientists have measured speed of light and concluded that the speed of light in a vacuum is finite and constant. But due to its enormous speed, all necessary observations have not still done.

Danish astronomer Ole Rømer first measured the speed of the light from the motion of the Jupiter's moon "Io". The Earth revolves in the orbit around the Sun, and the speed of orbiting by the Earth is greater than that of the Jupiter. So, in a different part of the year, at some time the Earth goes nearer to the Jupiter and at other part of the year the Earth goes farther from the Jupiter. Ole Rømer assumed that the Io must orbiting Jupiter with same speed whole the year, and there can't be a seasonal variation in the speed of orbiting of Io around the Jupiter. By timing the eclipses of the moon Io, he measured a speed of light and derived some conclusions.

His conclusion about the speed of light was, "The Speed of light is finite".

He concluded that, Eclipse was delayed when the Earth was away from the Jupiter. Means light don't reach to observer from the source instantaneously. Furthermore, his experiment also concluded that Light takes equal time whether the source of the light moving toward the observer or moving away from the observer. Observation of the binary star system also led to the same conclusion. It means that once the light enters the space, its motion related to Solar system becomes constant. The light from the object moving toward the earth doesn't move faster in the space and the light emitted by the object moving away from the earth doesn't move slower in the space in relation to earth.

So, conclusion is "speed of light is finite and constant in the vacuum."

Aberration of the star light.

We can imagine a situation in which a star is situated many light years away from the sun in a direction along with axis passing from sun's pole. Earth is revolving in the orbit around the Sun.

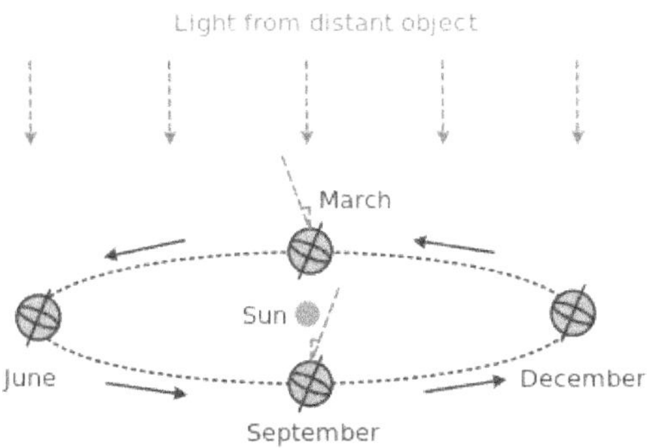

Figure 2.1

[Reference and Image credit: Wikipedia]

As in the figure 2.1 above, the star's apparent position changes according to the direction of the earth's motion. This phenomenon ruled out the existence of Aether moving along with the earth. It has been argued that if Aether was there getting dragged with the earth, then the light would also be dragged within Aether and there would not be stellar aberrations.

So, according to physics, this observation rules out the dragging of Aether within a vicinity of the earth.

Now only one possibility remained for the possible existence of Aether was as a stationary Aether which remained stationary and not getting dragged with the earth. If such stationary Aether was there, then it must have been moving in relation to the earth, because the earth moves in a circular orbit around the Sun, so it is not possible for Aether to remain stationary in relation to earth during whole year.

In such a case if we propagate the light on the earth, there must be a drag on its path due to relatively moving Aether. The speed of the Earth in the orbit around the Sun is approximately 30 km/sec, which is fast enough to produce a measurable drag effect on the light.

With this plan Michaelson and Morley performed an experiment with the apparatus made by Michelson known as the Interferometer.

The negative result of this experiment concluded two things.

1. No existence of stationary luminiferous Aether is there, because there is a no drag effect on light waves.
2. The speed of light is finite and constant for the observer when source is moving along with the observer.

Albert Einstein took all these results and observations in a consideration and concluded following conclusion,

1. The speed of light is constant for observers in all reference frames irrespective of the relative speed of the source of light.

2. Rejection for the existence of the substance Aether in a special theory of relativity

We can see that; Aether was rejected due to above explained observations. Light propagates with the constant speed in relation to the observer on the Earth, it means that, there is no moving medium around the Earth moving in relation to the Earth, which can drag the light along with it. This creates a relation between light speed and the observer. This is called observer specific constancy of speed of light.

So, these were the reasons in support of the assumption of Albert Einstein in the special theory of relativity.

If Einstein accepted the possibility of the existence of Aether, then that Aether can't be in a stationary position with reference to all moving observers, So, that couldn't comply with the second postulate of the special theory of relativity, an observer specific constancy of the light. So, he rejected the possibility of any type of substance as Aether medium that can be moved along with or in relation to the observer. In other words, he denied existence of anything in the vacuum space that can affect the light propagation.

This concludes, either Aether as a medium for light wave can exist there or a second postulate of the special theory of relativity can be true.

Note: (part of this chapter that include ole Romer's work, and aberration of light is taken from the Wikipedia, a public domain.)

3. Indirect support to Brahm

There is a no single General concept about the universal primary law that could define the foundation of the physics. Mass and vacuum still remained undefined fundamentally. Some theories deny existence of any type of Aether medium like the special theory of relativity, while some theories and observations are pointing toward some existence in a space filling the space and indicating that absolute space is not really absolute blank. Some parts of such theories are pointing toward Brahm.

Such theories or concepts are enumerated below.
1. Relativity.
2. Higg's field theory
3. Lorentz's Aether theory
4. Kinetic theory of Gravity, a push gravity model by Nicholas Fatio.
5. Quantum vacuum
6. Quantum entanglement
7. Young's Double slit experiment
8. Structure of the Galaxy
9. Hot core of the Earth
10. Magnetic field of earth and other rotating planets
11. Gravitational waves
12. Unsolved queries like inertia, mass, gravity, strong force

1. Relativity

The general theory of relativity indicates to Neo-Aether which was a late failed try by Professor Albert Einstein to re-introduce Aether. In the general theory of the relativity space is provided with the property to build geodesics which are the curves formed by masses around them in the material free space that can affect the motion of the mass within it and have a physical property that manifests Gravity. Gravity is explained as a curvature of the space-time matrix which create geodesics. Though Einstein tried to call it a neo-Aether and word like relativistic Aether has also been used for that Aether, but mainstream scientists did not accept it, because it was contradictory to the special theory of the relativity, and also against the conclusion of Michaelson-Morely experiment. But Logic says, if there is a nothing in the space how a nothing can be curved or folded to create a geodesic. This suggests a must existence of something in a space that forms gravity and governs time.

Gravity is conceptualised as a tendency of the masses to move toward one another. In the concept of four-dimensional space-time, mass is always believed to be in moving state. The stationary looking mass is moving in time dimension. When a mass advances through a time, the local direction of the time in a spacetime near a bigger mass is tilted to the centre of that mass. That creates a drag of gravity. Means the acceleration formed by gravity is progression of the mass on the geodesics tilted due to bending of spacetime by mass. In this concept, time is undefined property of the space. Mechanism by which mass creates curvature in surrounding space is not defined. This theory of relativity is incomplete in its concept at fundamental level, that point outs toward missing mechanism. This shows a room for Brahm and Aether. How steady masses keep moving in a time dimension is explained in a chapter 19. Reason for Time dilation and observer specific constancy of speed of light is also explained as a wave dynamic in a chapter 19. These two things also demand existence of Brahm including Aether.

The limit for the maximum speed of the matter in a space described in the special theory of relativity also points toward something restricting the speed of mass in a vacuum. This indicates some hidden existence in the space affecting the speed of the mass. If vacuum is a nothing, then there should not be any limit to the maximum relative speed of masses logically. When we gain the common sense from this theory of Brahm we would realise that if the vacuum is a nothing then there can't a stable mass be existed surrounding by the vacuum. Nothing can remain stable if surrounded by nothing.

2. Higg's field theory

In a modern physics Higg's field theory provides the explanation about the mass property of the particles. It is assumed in the higg's field theory that interactions of the particles with the higg's field are responsible for the mass property of the particles. Higgs boson is the fundamental particle associated with the Higg's field, a field that gives mass to other fundamental particles such as electrons and quarks. Particle physicist believes this field exists throughout the universe.

The Higg's field is a spinless field that fills the universe and has a nonzero vacuum expectation value. This means that the higg's field has a constant value everywhere. Higg's boson discovered in 2012, has no electrical charge, a short lifetime, and a spin of zero.

The higg's mechanism is the process by which the Higg's field gives mass to the quantum particles. The Higg's field breaks the symmetry of the electroweak interaction, causing the W and Z bosons to acquire mass. I don't know how it happens, there is a no source available on the internet that can explain this as an imaginable concept and not as a math. Other particles, like electrons and quarks, also acquire mass by interacting with the Higg's field. It is assumed that without the Higg's field, elementary particles would travel at the speed of the light, making it impossible for them to form atoms and molecules. Thus, according to above argument, this field is imagined providing only mass to the particles, but existence of particles doesn't require this field according to the Higg's field theory. This is different from the gravity forming flux in a Brahm which not just provide mass and inertia, but it is this gravity forming field of flux that forms particles. Without this flux no particle could sustain their existence. This flux also provides a field that is gets converted into a magnetic field.

Thus, Higg's field is the imaginary field that represents only resistance property of Brahm. Higg's field theory fails in explaining inertia. This theory proposes a concept that Higg's field provides a resistance to the motion of the particles, this resistance slows down the speed of the particles and thus provides inertia and mass. I don't think this can explain inertia. Because inertia is not just slowing down the speed of motion. It is a property which allows sustaining motion also. How a resistance could help in sustaining motion through field providing that resistance! So, such field which is not even categorised as fixed or moving don't comply with the nature of the inertia. But still acceptance of such field supports existence of something in a vacuum interacting with every particle.

3. Lorentz's Aether theory

Professor Hendric Antoon Lorentz introduced Lorentz Aether made up of the electron field in the 1904, but there were some mistakes in the calculations pointed out by Poincare. His theory was in favour of the stationary Aether medium but as we all know it lost the chance of improvement as the theory of special relativity rejecting any kind of Aether, superseded Lorenz's aether theory.

4. Kinetic theory of Gravity, a push gravity model by Nicholas Fatio.

Nicholas Fatio de Duillier was a fascinating and multi-faceted figure in the late 17th and early 18th centuries. He was mathematician, natural philosopher, astronomer, inventor and religious campaigner. He was a friend and close associate of the Sir Isac Newton. He was originator of the mechanism of gravity as a shadow gravity or a push gravity model. What could be better than a mechanical explanation! But This theory did not gain much attention of the scientists because scientists could not consider it fit for some observational findings; were actually it fits. In this theory absolute blank space was replaced by a gravitational environment which contained moving energy carrying corpuscles which were named by Nicolas Fatio as "Intramundane Particles". This theory gave insight that gravity is not an attraction but instead it is a push actually. This was very sensible. That was the most brilliant explanation of the gravity which is still being ignored. So, for explanation of the gravity as a mechanical push, one has to accept the existence of material things in a vacuum. That is a step toward Brahm. According to me, this was the incomplete theory representing Brahm at gross level.

5. Quantum vacuum

As the science advances, scientists find reach to the smaller scale more and more. A famous quote of the Quantum physicist Niels Bohr is "Everything we call real is made of things that cannot be regarded as real". This quote gives the insight into the reality that solids are not so solids and blanks are not so blanks.

When single mass particle is observed, it was not like a quite solid ball, but it was like a something carrying enormous energy and motion and in continuous contact with the surrounding to which it can transfer energy.

According to the present-day understanding of what is called the vacuum state or a quantum vacuum, it is "by no means a simple empty space". According to quantum mechanics, the vacuum state is not truly empty but instead contains fleeting electromagnetic waves and particles that pop into and out of the quantum field. This shows that vacuum has the potential that can absorb the energy or mass or provide the energy or the mass. This is called quantum fluctuations. These fluctuations involve the temporary creation and annihilation of particle-antiparticle pairs, known as a "virtual particles." Contrary to the classical idea of a vacuum as empty space, the quantum vacuum is understood to be a dynamic environment. It's not a state of nothingness, but rather the lowest energy state of a quantum field. This baseline level of energy is called zero-point energy. This means that even in the "emptiest" space, there is still inherent energy. The quantum vacuum is thought to play role in relation to dark energy. Some theories suggest that dark energy, which drives the accelerated expansion of the universe, originates from the energy of the quantum vacuum. The understanding of the quantum vacuum is based in the theories of quantum field theory. This theory is the framework in which the elementary particles and their interactions are studied.

These observational data are strong enough to accept the presence of something in the vacuum space that interacts continuously with matter.

6. Quantum entanglement

Scientists have found the relation between two entangled photons or particles, according to scientists that relation stayed even at far distance, this means that there is some connection between them even when they are physically apart from each other.

Quantum entanglement is a phenomenon in a quantum physics where two or more particles are connected in such a way that the state of one particle is dependent on the state of the other particles, even when they are separated by a large distance. This connection is believed to be instantaneous and seems to break a fundamental law of the universe. There is a recent claim of measurement of the speed of the quantum entanglement also. Albert Einstein famously called this phenomenon "a spooky action at a distance" because it seemed to contradict his demonstration that no information could travel faster than the speed of light. However still it is believed that quantum entanglement does not allow for the transmission of information at a speed faster-than-light. It is said so because scientists believe that even if the information is transmitted through entanglement, it couldn't be interpreted without communicating to the source who sent information. This communication must be at a speed below the speed of light, so human can't use quantum entanglement for information transfer with the speed more than light. But this doesn't mean that two quantumly entangle particle don't interact with each other with speed more than light. Actually, this connection is against the prediction of special relativity.

Though I am still struggling to understand quantum entanglement clearly because of poor sources, I have doubt about any such connection. It is easy to believe that two particles can be entangled with each other because of the same source and time of origin. But continuous connection because they were entangled is not digestible. Heisenberg's uncertainty is due to lack of tool that can measure quantum particles without too much disturbing them. But applying that uncertainty principle to quantumly entangled particles by assuming connection between them even at very far distance is something weired. EPR paradox demands hidden variables that allows entanglement and specific behaviour in certain interactions. Ruling out hidden variables by testing method like bell test won the Nobel prize. But what if 30 percents of the photon believed to be entangled were not actually properly entangled! It will yield this type of result indeed. I would not like to believe in quantum entanglement until there could be a direct measurement of the entanglement.

But, if it is true that distant quantum particles can interact with the speed more than light, then what could be the medium? Brahm is a such complex thing that can provide the answer. Speed of mass in Brahm is not restricted to the speed of light if that mass is smaller than Aether units of our scale. As the scale becomes smaller relative speed becomes more intense to provide gravity push. If electrons move with speed near to the speed of light, particles stabilizing it with push gravity must have speed many times that of electrons. Same thing is true for the photon if it is a particle. Actually, smaller scale particles are not very fast, but larger masses are slower relative to them. Time perception is due to motions, that's why slowing down of time is not perceptible at a scale with larger masses. You can check details on this in a chapter "Scales".

7. Young's Double slit experiment

This experiment shows interference patterns when electrons are fired along with double slits toward detectors. It provides a conclusion that even particles have a wavelike property. When electrons were fired one by one, even in that situation, that showed an interference pattern. There can be one of the two conclusions possible for this phenomenon. 1. they interfered with themselves by moving through both slits simultaneously 2. they were doing something mysterious. This is the unexplained phenomena that can be explained by accepting Brahm. Electrons carry a magnetic field in Brahm, though electron pass through only one slit the magnetic field of the electron passes through other slit, there may be an interference between two magnetic fields passing through slits, that creates a summation and drag, and it deviates the path of electrons and creates interference pattern on the detector screen. When you disturb environment by detectors, it wouldn't be able to bring same effect.

8. Structure of the Galaxy

According to modern science, Galaxies are unstable structures theoretically. The mass they carries is not enough to create required attractive force of gravity to hold the whole structure of the galaxy in a rotating motion with observed speed. According to modern physics mass was lacking in all galaxies. There was an imagination of Dark matter to fulfil this lacking proportion of mass. Recent studies revealed that the lacking mass is found as a cloud of thin gases distributed around galaxies in a glob like area. This finding rule outs the need of dark matter.

Distribution of mass in this pattern favours the effect of the structure like a Torus. Galactic centres are also a mystery and assumed as a blackhole. That can be explained in a better way with mechanical push gravity, as explained in the respected chapter. Shape of Galaxy and distribution of gases around it can be explained by the topic Torus formation described in the chapter 5 of the part 2 of this book.

9. Hot core of the Earth

It is assumed that core of the earth is hot, because of its previous hot state in the past of the Earth's history. This book explains this as an essential acquired state due to consequences of push gravity in a Brahm. The larger the stable celestial body, hotter the body. Core is hotter than outer crust. These fits well in the consequences of push gravity which is a Brahm gravity model.

10. Magnetic field of the earth and other rotating planets

Torus formation and magnetic body explained in this theory can explains the origin of the magnetic field of the earth. Combined magnetic field of few magnetic bodies further aligns more magnetic bodies and creates more powerful magnetic field with increased range. Axial rotation of the earth doesn't allow alignment in other directions, and it just allows Magnetic fields summation in direction nearly parallel to rotational axis only.

11. Gravitational waves

Waves are serial fluctuation in a medium. Gravitational waves exist, means a medium is there.

12. Unsolved queries like inertia, mass, gravity, strong force

Explanation of these things are incomplete in absence of Brahm. See relevant chapters for explanations.

4. Possible nature of Brahm

Let's derive some primary properties of Brahm.

We would like to understand the nature and characteristics of Brahm by understanding its interaction and expressions in relation to observable things.

At first, there is an explanation of Brahm in relation to "Light"

We have seen in previous chapters that the observed behaviour of light in space is weired as a wave. It has still not been decided whether the light is a wave or particle.

Light as a wave: - If we want to represent the light as a wave in a Brahm, then we should compare the properties of the light as a wave in Aether medium with some other wave in other medium.

Let's assume a water as a medium and wave in a water as a wave of the medium. Now we can see that if we create a fluctuation on a water surface then it travels as a wave form, which disperses or spreads in a water. It is not possible for the wave on a water surface to travel in one direction without being spread in all directions. Waves in water are dispersing waves. Same as sound waves, it is also dispersing waves. What are other waves? We don't know many types of waves in a continuously distributed medium. Whether they are longitudinal waves or horizontal waves, invariably they get dispersed in a medium during their propagation. Do we know about any non-dispersing waves? Non dispersing waves are waves which travels in one direction and during their travel, they don't get dispersed in surroundings.

Do we know about the sound that can travel in one direction only like a light? No, no such thing is possible, sound always gets dispersed.

But what about the light in vacuum? Light travels in a single direction. And for a very long distance and time it remains unchanged in its shape and energy. It doesn't get dispersed in vacuum. If we transmit a light in one direction, that goes in that direction and unlike a sound or water waves it doesn't get dispersed in all directions at least up to very far distance. Even if we assume Aether as a medium for the light, it is hard to imagine a wave within it which don't get dispersed in surroundings! Is it possible to create such a wave in any medium? The answer from experimental physics is, they don't know, and Scientists haven't performed any experiment that can give insight on this matter.

What about a wave in the elastic medium? Does a wave in the rubber sheet can travels in single direction? The answer is no. so how is it even possible to imagine a wave like a light and a medium that can propagate it as observed? If rubber block is very huge and a fluctuation is there as a wave that can travel in a single direction. So, just airlike or a waterlike property of Aether doesn't fit in these requirements. The medium must be of some different kind which made possible the existence of non-dispersing waves within it. It must have elasticity and resilience.

Now looking at another problem, as we know that wave moves as the medium moves. It has momentum same as that of medium. Waves don't have an independent momentum that doesn't affected by momentum of the medium. Means it is not possible for wave to remain at specific place in a reference frame when medium start moving relative to that reference frame. The wave moves with the medium. So, if Aether is present stationary to the earth within vicinity of the earth to some extent in the space than stellar aberration should not be there, and if Aether medium is stationary to the sun and moving in relation to earth than Michelson Morley experiment should show fringe shifts. Does it mean that Aether as a medium for light can't be possible? Role of tides are described in the topic Inertia in the part 2 of this book and light propagation also explain in the chapter "Light".

In another chapter possibility is shown about the light as a massive object also. So, stand is not taken about light as a wave or a massive object, but possibility of existence of non-dispersing wave and fish projectile is shown in the theory.

Let's see the next point about light.

The wave must decay in a medium because as it travels in a medium it loses energy in the surrounding and thus get dispersed and progressively disappears. But we don't observe decay in a light wave. We don't know about the maximum distance after what light disappears as a visible light.

How a wave can preserve its structure and characters for such a long distance and such a long time! If light is a wave and not massive particles, then a medium propagating it must have special property that can carry such wave without its dispersion in surrounding. Light as a mass can also fit in this observation.

The next challenge to Brahm is resistance to the moving mass. As I have described Aether in this book like a voluminous substance similar to liquid or solid and not like a very thin gas, how is it possible for it to not provide a significant resistance to the moving mass? If we assume inertia as a resistance by Aether in a space, then it should remain present continuously and must stop the moving object with providing a negative acceleration. But we can observe that resistance is not seen once a mass achieves velocity in space as a characteristic of inertia, instead there must be a need of force to stop that object from moving in the space. So, we don't see the expected continuous resistance to the masses. It is very interesting to explain this point in a topic inertia of the chapter 6 in a part 2 of this book. But I want to clear here that it is not true that Aether and Brahm don't provide any resistance to moving masses. In fact, Mass is formed by complexly arranged multiple vortices of waves in Aether itself. So mass is a curved, whirled aethereal waves and this complex moves as a wave in Aether. So, there is a no separated ether between masses, but mass is complex of aethereal waves. So, how an Aether could provide resistance to masses when it actually propagates them! Still there is a limit on the speed of waves within Aether, and that is applicable here. At particular speed all waves travel within Aether. Because mass is a complex wave it has some circular motions of its units and some linear motion. Sum of both may always remain constant. This is a reason there is a no Aether drag and resistance to moving objects. So, the wave with only linear motion has the fastest speed. Light may or may not be this thing. Inertia is due to gravity forming flux. Its Balance decides relative motion of mass within Aether explained in a topic inertia.

Let's understand Aether in relation to gravity. You will find in upcoming chapters in the part 2 that, this Brahm theory has acquired a push gravity model to explain gravitational force. This essentially allocates a property of gravitational environment to Brahm.

Internal relative motion of Aether:

1. If Internal relative motion of Aether is possible with change in the arrangement of lattice units, it could be termed as a fluid like Aether.

    In this case primary mass may be a vortex of Aether fluid within Aether. It is not a wave, but a vortex. Here current and vortex in Aether are accepted as possible things. A vortex with torus formation and energy field woven with torus can behave as a resistance within Aether that represents a mass which is stabilized by push gravity. Here push-gravity can be formed by both aethereal vortices and aethereal waves.

2. Internal relative motion of Aether units is possible only without change in the arrangement of lattice units (Elastic solid like Aether).

    In this case the primary mass is a couple-waves within Aether whirling in a circle. Gravity forming flux pushes such primary masses together and forms a vortex of waves within Aether. This whole vortex is a complex of wave-couples and wave clusters. Here aethereal units don't move freely but can only oscillate within some range. Here also the vortex of waves in Aether with torus formation and energy field woven with torus can behave as a resistance within Aether that represents a mass which is stabilized by push gravity. This concept gives rise to the concept of Time-dilation which is there due to wave nature of the mass.

About Aether and Brahm.

- It is **luminiferous.**
- Gravity is a phenomenon occurring in Brahm. So, Brahm has a **gravitational environment.**
- Energy is either a "moving mass" or a "wave" in Aether, in short it is a motion carrying momentum. So, Aether has property that **it can carry energy as motions.**
- It is not perceptible until it gets expressed in some form like wave or mass. So, it is **transparent, translucent,** in short, in its pure form it is mostly **impersonal.**

So, here I have represented some properties of Brahm. It is better to describe each property in detail as separate topics in this book. Chapter 2 and 3 in second part also focused to highlight qualities of Brahm. So, these were the few of many properties of "Brahm including Aether."

Description below should be read after completing this book to grasp the meaning. Still, I have discussed it here because of its relevance to the topic of this chapter.

Brahm and Aether are not same but still they are same. When I talk about Brahm, it is composed of moving masses and waves creating push gravity environment. That forms vortices of masses. That forms magnetic fields. That stabilizes particles. That creates various forces. This dynamic masses and waves with Aether together called Brahm. When this gravitational environment forms a symmetrically arranged units that form a continuous substance, that substance is called Aether. Once this substance is formed, it behaves as a bearer medium. Vortices of waves in this substance forms primary mass above the scale from that Aether units. That primary masses can have motions. They act as particles forming push gravity to larger masses. Waves may also contribute. All these together again form Brahm at that scale. So, Brahm forms 3d lattice Aether and Aether with its dynamics forms Brahm. In an aethereal world there could not be a Brahm without Aether, and there could not be an Aether without Brahm. These two together form Existence or Nature. These are just a combined expression of mass and blank space, means combined expression of something and nothing. Still, I have used two terms to describe the Nature because Brahm represents a compulsory dynamic state that is essential to form Aether. Also, there is a one possibility exists that may include a Brahm without Aether as described in last chapter. That's why both term Aether and Brahm are used.

Nothing is a space where mass or something can move. We can see that cluster of vortices in Aether can move within Aether. So, for the cluster of vortices, Aether behaves like conventional nothing or a space. So, there is no need for absolute nothing to get the space. But without multi scaled nothing Aether can't be flexible. Aether is well arranged masses of the lower scale Brahm, means flexibility of this scaled Aether is not due to absolute blanks but it is due to relative space provided by Brahm and Aether at lower scale. So, if we want to find absolute nothing, we have to go inner and inner to smaller scales. Similarly, if we want to find absolute solids, we have to move inner and inner scales. What we could find is Aether and its expressed part as a Brahm, but not absolute solid or absolute blank. That means absolute blank and absolute solid may only possible beyond infinite number of smaller scales. Still, I have assumed absolute solid and absolute blank in a chapter Honi to understand ultimate possible distributions of masses in the space and consequences that leads to Brahm and Aether.

End of the part I

Part II

Brahm model of the universe and fundamental Physics

Introduction

This part explains the existence as two basics. One is a Brahm that require two fundamental things, a blank space and a solid mass. If we accept these two fundamental things as a self-stable than there is a no need of scaled Brahm with Aether.

But if we describe a primal solid as a vortex of waves within a substance Aether then this Aether becomes an only supreme fundamental existence that can form Brahm and physical world. But if we inquire this Aether for its roots, then it would be found that it is a manifestation within a Brahm powered by kinetic energy and made up of primal masses in a space. Further root of those masses would be vortices of waves in a finer Aether. This is an infinite sequence in the existence. Aether forms Brahm and Brahm forms higher scaled Aether. So, we can't get the rid about primal fundamentals, but we can percept the structure of the Matrix.

In this second part, physics is explained as a model based on Brahm including Aether. There is no direct assumption of Aether or Brahm for the sole purpose of the theory, but it is explained that there is an essential need of Brahm to explain the real world in a realistic approach. If we try to neglect Brahm and try to explain the reality, then the world becomes miraculous, common sense get killed and nothing remains as logic to differentiate between the real and non-real. So, to understand the world as a real we need to accept that Brahm must be there, and everything is there as manifestations within Brahm. *So, key point is a realistic mechanical approach and no miracle strategy.* The meaning of the word logical is "Whatever could be explained with the continuity of logic with mechanical concepts based on the fundamental foundation of the theory". I also see the relation of the word "Real" with the word aethereal. Means Aether is beyond real and manifestations within this Aether which could be expressed as aethereal and thus Real. By accepting Brahm with a simplicity as much as possible, we can avoid many assumptions of modern physics like extra spatial dimensions, singularity, limitless vicious gravity and relativistic blackholes, non-unifying forces, mysterious gravitational force working without any mechanical interactions, mysterious strong and weak forces, some weired properties of light, different distinct fields for different effects like quantum field, higg's field, electromagnetic field etc. This theory is imagined as whole concept, for the existence of universe as a whole and try has been made to incorporate as much facts as possible within the framework of the theory. Still some aspects could not be explained with good accuracy, like an exact quantified calculations of subatomic particle interactions or anatomy of the quantum particles. These aspects need further development of the theory with incorporation of data based on experimental science. That works needs précised data and accurate equations to relate phenomena. This theory provides a framework, and Science has to freshly discover everything within this framework, if this theory gains an importance.

In the first chapter we will see, why it is must to accept the existence of Brahm and there is a no other choice remains if we want to keep the existence of the world as a real and logically explainable. As soon as we accept the existence of Brahm, we would be able to derive properties of Brahm by using logic with available data about facts and thus building a concept of the real world which explains most of the things in a realistic manner. Brahm can contain a mechanical world, where everything is nothing but a mechanics or dynamics. We live in a real world and have data as observational findings in this world, so, in a process of logical explanations in this book, sometimes there may be a neglection of some possibilities without thinking much of it if that possibility is not deriving results on the way to the real world. Logic and imagination will guide us on the way to derivation of the gravity, derivation of the mass, etc. In succeeding chapters, there is an explanation of the phenomena between masses, explanation of the light and its properties. All this based on the thought experiments and logical co-relations only, so quantitative data is not there, but logic used is so simple and mechanical in a nature that a common person with a simple common sense can grasp the essence of this model. Simple minor mistake in application of math can lead to miraculous non real outcomes with paradoxes. So, it is not advisable to leave a common sense for a math in the formation of overview of the model. This overview of the model presented in this book doesn't contains any math. Further development of the theory by incorporating maths for precise predictability is a pending work for now for this model. Math is used to describe Relativity in relation to this theory. One can compare Maxwell's equations in this framework also. One can apply math for shadow gravity model also. The track of the model doesn't follow the way to modern science theories, and it may be in support of some less known or relatively ignored or discarded theories. For the development of this model, I didn't consider any theory as more important theory or proved theory. It is this model that decided during its development that which theories should be incorporated and which should be rejected or neglected.

Dr. Hasmukh Devidas Rathod.

52

0. Thinking.

This part is just an additional extra chapter, which has no direct relation with the theory of Brahm, this is added just for fun; you may skip this "0" chapter.

Sense, Logic and other words related to thinking

Sense: sense is the ability to percept and understand anything.

In humans, it is a process of getting data, processing & interpreting those data and realizing about things related to those data. These include involvement of sensory system, processing and analytic activity of brain and cognitive system of an individual.

Intellectuality or intelligence: It is an ability to use logic and data to understand something that may result in some conclusions.

Logic: An ability to arrange data in a way that must obeys laws in a way that can create a meaning is called a logic or a reasoning.

According to no-one but me, origin of word logic is Law-mosaic.

Knowledge:

A logically well-arranged data is called knowledge. Knowledge may be true or false.

Common sense: When observer observes a system he may found some patterns. There are events that happen in a particular way only and no exceptions are there. That particular way is considered as a law. Understanding such laws forms a common sense that provides a sense to think about possibilities and mechanisms. Such sense based on the Laws is called the sense. When without understanding laws, a statistical sense develops on the basis of superficial analysis of the system by considering only few common events as unchangeable facts it is called a common sense.

Conditioned sense or common sense:

When laws accepted in a common sense are accepted without further understanding of the phenomena behind those laws which could create a mechanical sense than it is termed as a conditioned sense or a common sense.

Philosophy (Pûrân):

When unobserved area (gap) in knowledge is filled by linking known portions by pure logically created filling portions that may help to complete the picture by preserving continuity with some knowledge in a way that could create a sense, the process is called philosophy.

Means, philosophy is an extension of the knowledge created with linking-logic and imaginations in absence of required necessary observed data that fills gap inside knowledge and thus provide a larger picture of reality which is not a whole confirmed truth. So, it is less valuable than pure knowledge. It may be true or may be false. It may misguide to impossible consequences also.

Few types of thinking

Overthinking:

Creating an imaginary building of concepts on the foundation of any base without use of enough observed data and proper sound logic is called over thinking. Inability to rule out impossible possibilities by using proper logic and data leads to overthinking. This may lead to paranoid nature.

0. Thinking.

Critical thinking:

Re-thinking by using logic at more fundamental level by rejecting superficial common sense and identifying easily missed logical errors is called critical thinking.

This helps in reanalysis.

Determinism:

Overweighing one or more possibilities over others powered by personal beliefs or emotions is called determinism.

Prejudiced thinking:

Thinking with determinism with own beliefs beyond its applicability is called prejudiced thinking.

Examples are:

1. Thinking that everything must have a starting and end.
2. Anything can come into existence only if it is actively created by someone.
3. Same type of things must follow the same protocols.
   i.e. I) All guns must have wooden shoulder-rests.
       ii)Smoke is there, there must be a fire somewhere.

Restricted thinking (kup mandukta)

Applying local laws universally or not applying universal laws locally is called restricted thinking.
i.e.
A crow must be black.
A peacock must have colourful feathers.
A Material must be an element.
Existence is the universe and universe is the existence.

Conflictive thinking:

Accepting two or more contradictory possibilities simultaneously without realizing its contradictions is called conflictive thinking.

Strict thinking:

Avoiding acceptance of contradictory possibilities simultaneously and rejecting each due to their internal contradiction is called strict thinking.

Liberal thinking:

Giving equal weightage to two or more contradictory possibilities based on its individual logical acceptance without being deterministic is called liberal thinking.

Analytic thinking:

Checking any concept with multiple tools and checking its multiple aspects in relation to established facts is called analytic thinking. This is a tool for critical thinking.

Abstract thinking:

Finding similar patterns in apparently unrelated things and trying to find relations between them using such patterns is called abstract thinking.

Divergent thinking:

Instead of finding solution on the mainstream way, trying to find alternative less appropriate way is called divergent thinking or divergent attitude. Habit to think routinely on this way shows a shunning attitude of the person for responsibilities.

Ignorant thinking:

Thinking without use of proper logic, laws and facts.

Laws:

Analytic intellectuality is an ability to identify normal consequences and to distinguish abnormal consequences. When an observer with analytic logic power and a memory observes a system for enough time repetitively, the observer understands the behaviour of system by identifying patterns, repetition of same or similar consequences which he categorizes as normal findings. In such findings some findings or phenomena could be categorized as primary and basic phenomena which remains always true. Such always consistent phenomena are called primary laws. After such laws could be derived, that can help in understanding other complex phenomena as consequence governed by these primary laws. So, sense is also to be considered as an ability to understand complex phenomena with the help of few primary laws.

A sense that can differentiate between possible and impossible phenomena by calculation based on Laws, data and the logic power is called a common sense. Common sense can create an event in an imagination with sense of its possible possibility based on logic. Thus, common sense provides a power to differentiate a real and non-real to some extent. The possibility is always there that what are considered as Laws are not only laws and there may be some hidden laws and existences that can influence the result. That's why common sense may lead to different out-come away from the truth.

Some laws are primary laws and others are secondary or derived laws. Sometimes we mistaken secondary laws as a primary law due to its influence in a routine. The common sense that accepts derived laws (phenomenal effect) as a primary law is called conditioned common sense. So, our routine common sense is a conditioned common sense that don't allow us to see detailed image of the reality. Identifying primary phenomena and root cause behind secondary laws with deeper sense can lead to pure unconditioned common sense, which can help in understanding about unexplored regions of knowledge field, and it can predict some missing mechanisms. I would like to explain this with one example.

Attraction:

Some non-existing phenomena or things have a definition due to conditioned common-sense. i.e. "Attraction". Attraction is thought to be a primary force working on two objects in direction toward each-other on the line joining them and thus pull two objects together. Means it is like an effect of invisible elastic bands functioning itself attached to two objects that can create an attraction force between them. If two objects move toward each other by force acting upon outer side then we don't consider it as the attraction force, but we would call it a push or thrust. Science have not yet confirmed, if true pure attraction can exist or not. We know some attractions like gravity, magnetism, electro-magnetism that can move two objects toward each other. This conditioned our common sense in a way that we accept that there can be a pure attraction and these are the examples. But if we inquire these phenomena in relation to mechanical concept as a pure attraction, we will require the interaction that must behave exactly opposite to the newton's third law of motion. If newton's third law is true for material body and energy, then opposite can't be true. If opposite is true, then it breaks newtons third law of motion. This leads us to inquire the phenomena excepted as a primary attraction again. We have to look for, if there could be a hidden complex mechanism there that works without breaking newton's third law and creates an illusion resembling a pure attraction? There is a Something in favour of this way of thinking. Maxwell's equation draws our attention to a law, "there could not be an isolated single pole of a magnet". If we break magnet in two, each of both magnets will have two poles. There is always a field surrounding a magnet that works on the portions facing external sides. We can harvest energy by moving a tangle of wire in a magnetic field and that don't decrease the strength of magnetic field abruptly. If that magnetic field had the energy due to momentum of something that was a part of the magnet, then its momentum must have been limited and that should have been resulted in an abrupt decrease in the power of the magnetic field after harvesting energy from the magnetic field by moving wire tangle. This clear out that magnetic field is not only related to magnet itself, but it is connected to infinite amount of source from the surrounding. You can compare magnet with an assembly that can create a deviation of the water flow that can create a whirlpool in a flow. Attraction between two opposite poles of the magnet is not a separable interaction between two opposite poles only. Possibility is there that it can be a complex interaction between two complex things with involvement of surroundings. Similarly, gravity is a mystery in a modern science. Possibility of push gravity can't be ruled out. So, theoretically we don't have any base in support of pure attraction. There is no thinkable conceptual mechanism in support of pure attraction, this just rules out its existence until we could get some contradictory findings. These interpretation gives a firm argument that even chemical bond, elastic recoil of spring, pull by rubber band or springs are also pulls due to indirect pushing or thrusts. Means in real those systems create complex modified force that seems to behave like a pull, that is not a pull but a result of push and strikes forming thrusts at fundamental levels. You may imagine pulling of vehicle with rope. If vehicle is pulled by a rope tied to a hook on the front part of the chassis of the vehicle, then it is an example of the pull. But we can see here that, a rope transmits its force on the backside surface of the hook and that is a push. Now still we can argue that a hook pulls a vehicle because it is attached to vehicle on front side and not transmitting its force as a push from backside part. Well, this integration between hook and vehicle is from integrity of mass or metals. This integrity is a result of chemical bonds between molecules or atoms. That chemical bonds keep integrity of metal. Integrity of chemical bond is there due to structural integrity of atoms. If gravity, magnetic force, strong force is not there then this integrity can't be there. If these fundamental forces can be explained as an effect of strikes at fundamental levels than it could be concludes that the pull can be possible due to support to structural integrity by surrounding push forces. Means we think that a rope has tensile strength, but on the way of above thinking we can conceptualize that this tensile strength is provided by fundamental forces which are not fundamental really but an effect of fundamental level strikes or pushing interactions. Pulling is just changing equilibrium of inertia resulting in motion.

So, attraction and pull can be an illusion appearing due to hidden pure mechanical interactions following dynamics and newton's third law of motion. If someone want to explain pure attraction than one has to show a mechanism that can break newtons third law of motion, or one has to provide a mechanical model similar to model of action of hypothetical gravitons or gluons. The working model of gluon demands flux tube which further demands tensile strength and elasticity. Imagining flux tube between quarks with gluons inside it is surprisingly imagining an elastic thing with material finer than smallest fundamental particle with fifth fundamental force providing elasticity and integrity to that flux tube. If it is allowed to imagine such flux tube with gluon within itself between quarks then what is the need of gluons? Flux tubes with elasticity themselves can apply pull between quarks! The force which provides elasticity to flux tube can be considered as a new mystery and fifth fundamental force. So, actually gluon mechanism is a paradoxical impossible mechanism. Same thing is applicable in the case of graviton also.

1. Brahm, Aether & Requirement of Brahm.

Sir Isac newton once stated that, "You must have to start with something undefined". Actually, he might be talking about an essential need of a primary assumption which can't be explained further with the same theory or other theories. Or he might mean, "No one can find an absolute difference between absolute something and absolute nothing". This simply also means that "we can't completely understand the most fundamental existence anyhow".

There are no primary assumptions in this theory which is not explained further with the same theory. This theory is a self-stable and no undefined assumptions are there. Not a single assumption is there except a law of thermodynamics and newton's laws of dynamics are taken for granted. This means that this theory uses classical mechanics. Theory starts with nothing and moves over something, then gradually thinking about the possible stable state of that something, with the help of logical thought experiments. It is explained step by step that, "that something" can only be Brahm and nothing else. It is showed in a chapter "Honi" that, however you may imagine any existence as a something, as a mass, as a point, as anything, that would achieve stability only as Brahm if it could create a real world. Infinite repetition of the same explanation is the key that solves the puzzle of undefined something. But still INFINITY is the puzzle itself. The Nature of existence is defined in this theory, but with open ends on smaller and larger scales, with open ends for infinite past and infinite future also. If we call "these" to the structure of Brahm at particular scale, then we could say that "These and these inside and inside, continue for infinity is the answer. Only this could be the final answer, or could it be more precise?

Overview of Brahm World

Two possibilities about Brahm are Shown in this book. 1. Brahm including Aether. 2. Brahm without aether with undefined mass and emptiness.

First possibility is described in whole book and second possibility is described in brief only in last chapter of this book.

About First possibility

Aether is noble elastic substance, which can have waves and vortices of waves within itself. Aether with vortices and energy flux in itself is referred as Brahm in this book. Brahm contains manifestation of Aether including unmanifested Aether itself. For example, water in an ocean is a material, but ocean itself includes waves of water, currents in water also. So, term ocean includes water and its manifestations also. Similarly, Aether is a material or substance and Brahm includes Aether with its manifestations also including moving corpuscles, steady masses and waves. Mass is nothing but a complexly arranged vortices or whirlpools of waves in Aether which is stabilised in a state by moving energy carrying things called flux which forms push gravity. Aether is a material, or a substance made up with complex arrangement of vortices, mostly and stable in relation to one another. Stable units of Aether are formed by these vortices and ultimately these vortices are formed due to push gravity by moving masses and waves. These moving masses and waves forming flux are part of Brahm. So, Aether itself contains Brahm at smaller scale within it. That Brahm stabilises units of Aether and thus stabilizes Aether itself. But this Aether doesn't necessarily form vortices in its substance to form corpuscles and dynamics above its scale. Aether can have a mass in it and only mass can create an Aether. Both are interdependent and together creates a material world with Aether.

Aether: It is described here as an elastic substance medium that has units as stable complex corpuscles with fields surrounding it. These units form a medium with elasticity and resilience. That can carry waves, that can have vortices and whirlpools within it. Vortices are made up of masses. Primary mass is assumed as a couple waves or cluster waves. Thus, ultimately all vortices are made up of waves within Aether. There is a possibility of fluid Aether and fluid current loop as a primary mass within Aether but wave-couple as a primary mass explains time dilations. Still both possibilities are kept open either aether as an elastic medium or a fluid medium, but in this book preferably a mass as a complex wave instead of loop current is explained and Aether is described as an elastic substance with compressibility and resilience instead of a liquid with uncompressible nature.

Brahm: This includes Aether with its dynamics. Corpuscles formed by complex arrangement of whirlpools of wave clusters within Aether are part of Brahm. These moving corpuscles and waves together form a push gravity and Aether carries a gravitational environment. That Aether with gravitational environment is called a Brahm. If we imagine Aether with silent environment with no relative motion of its units, then it is an Aether but not a Brahm. When Aether becomes Brahm, it carries dynamics that can form structures which further forms mass and material world. So real existence includes Brahm with Aether, masses and waves.

If we imagine an Aether without mass like an empty space, even then that space is capable of carrying waves and is made up of material not expressing itself as a mass at surface scale (surface scale is a 3D and made up of units of Aether). These Aether units also need a support of push gravity to remain formed and stable. Only Material bodies or waves with momentum, moving basically as waves of smaller scaled Aether can create and stabilize units of gross scaled Aether. Means silent Aether is silent above its scale on a surface (3d), but at smaller scale or inner side that silent Aether is a result of dynamics of Brahm. Without those dynamics of Brahm, units of Aether can't have a stable existence. Means Aether is not possible without vortices or moving masses in it. If in a thought experiment, all vortices disappear all of a sudden, then nothing remains at observable limits and there is a continuous disappearance of existence would occur toward smaller and smaller scales up to infinity. This means Aether or any existence is due to dynamics only. Aether with vortices in motions can create the universe with simple mechanical laws. Moving things forms push gravity and other effects that forms clusters and structures and ultimately forms Aether itself. If there is a no motion, no vortices or whirlpools, then nothing can be there which can move and create gravitational push and further formation of structures and reality. In other words, everything as an existence is just there due to dynamics within a special magneto-gravitational infinitely scaled substance called Aether.

But if we consider Aether and Brahm as reality, and imaging a no motion state, then it could be possible at surface scale only. There is a law called conservation of the momentum. So, when we want motion free one scale in an Aether or at absolute motion free existence, then we have to find the place where all momentum can be accommodated. This simply means that there could not be an absolute motion free state of existence, but it may be possible that all momentum enters in smaller or inner scale and at surface within Aether there is a no motion or vortices and it may acquire a calm "Nirav" state for relative motion of units at that scale. As a result of continuous process when motion again enters the described scale, means at a surface within Aether, then it again forms a scale of Brahm and existence of real or aethereal structures at the level of its surface scale becomes possible. This word also provides an insight that aethereal is real. In a Brahm Aethereal is real part of Aether. What is manifested as physical existence either as masses or waves is Aethereal and aethereal is real. Unmanifested Aether is not real for manifested material world according to this scale restricted definition, but it exists, and it is real in casual sense. Why Aether is also Real! This is because Aether represents only one scale within Brahm, and it is itself a manifestation in a finer Aether. So, Aether is also aethereal and real. The word "Param" represents the state of Aether in relation to material world as the existence beyond material world.

What could be the first existence? A mass and emptiness or a substance precipitating that mass as a cluster of wave-vortices and providing space for motion? In a chapter Honi, a state is explained in which unstable mass is imagined there for disintegration in smaller fragments, and only stabilised masses by push gravity can remain stable as a mass. The process of disintegration of unstable mass is spontaneous and continuous, powered by laws of thermodynamics. It simply means mass, space and Aether are simultaneous existence with energy(motion) within themselves. None of these four can exist independently with absence of any of them. So, A trio of mass, motion and emptiness are described as a primary existence, but these all can exist as a manifestation of one entity Aether. So, Aether is made up of these Trio only and these trio is a manifestation in Aether. These all together with infinite scales are together called Brahm. These trio and Aether are not granted as primary assumptions, but they are showed as a result of all possible possibilities in a chapter "Honi".

Requirement of Brahm with or without Aether.

For the existence of Nature, requirement of Brahm is must, and no theory can explain the real universe as real without accepting Brahm by using unconditioned common sense. Why and how it is so, is explained below.

Suppose no Aether is there and space between masses is a blank space devoid of any real material thing, then there would be following problems.

1. Attraction becomes miraculous without push gravity environment of Brahm.

We have noticed an attraction since long time. So, we have accepted it as a primary force or phenomenon whenever we think of the matter in a space. The state of common sense is set at pure primary level for this model, that don't accept any effect without its mechanically explainable mechanism. Within a scope of this common sense there could not be any attraction at all. So, there could not be any pure attractive binding force responsible for integrity of mass because such thing is not just impossible but even unimaginable with preserving laws of classical mechanics.

The Gravity or any pure attraction without a surrounding or in between fields becomes a mysterious primary force without Brahm and Aether. Without Brahm, Gravity becomes a force that acts through the blank space and without mechanical contacts between masses. Even graviton or Gluon like mechanism, if imagined then it must behave against Newton's third law of the motion. Third law states that when two thing collides and if there is a no structural damage to that two things then they will move in opposite direction after strikes at an angle determined by Nature of the collision. But what gravitons do? They bring two things nearer to each other by having strikes with them one by one. Whatever interaction imagined here; it goes against the Newton's third law of motion which is a primary base for the commonsense. This breaks a logic of the mechanical common sense. So, graviton doesn't make sense. Mechanical Common sense and logic used in this Brahm theory don't accept that there can be an attraction at all. Yes, as mentioned in previous chapter this pure common sense doesn't accept the phenomenon called attraction. Why it is so is explained below.

Imagine, there are two solids with firm stable structure. Both present at some distance from each other. There is a third party which moves between them and collides with each of them one by one in a line crossing all threes. Let's use a name "Bounce" for this third party. There is an event of collision between one of those two firm masses and Bounce. Few possibilities could be imagined here, in an event where momentum is conserved.

First imagining a graviton mechanism. This concept is already rejected in a description ahead in this chapter

Second imagining Bounce moving and striking with two massive particles.

i). Bounce collides with one firm mass and moves back but with a net final momentum less then compared to previous momentum. In such condition Bounce will become relatively sluggish in motion gradually after many strikes and two firm solids can only get pushed away, because they have gained momentum in a direction of motion of Bounce while strikes. This preserves mechanical commonsense.

ii). Bounce collides with one firm mass and moves apart but with a net final momentum equal to previous momentum. In such condition Bounce will continue moving and colliding both firm solids and two firm solids will neither gain or lose momentum and remain steady there. This is against a mechanical commonsense and this doesn't help in describing a mechanism of pure attraction.

iii). Bounce gives momentum to the first solid with which it collides but in direction opposite to its direction of motion before collision. After collision both moves in a opposite direction toward the second solid. If this is true, then there is a murder of the mechanical sense and laws of classical mechanics. This behaviour is against the commonsense of classical mechanics including Newtons third Law of motion. Here primary law of Collison becomes "when smaller moving solid strikes with a steady bigger heavy solid, a smaller solid moves back after strike and also brings bigger solid in that direction of motion against newton's third law of motion"

We know that universe don't works like option two or three. We know that law of conservation of momentum is a primary law and that forms a sense of mechanics. This is why because this is the only way on which energy conservation or motion conservation could be possible. If Nature don't behave like this preserving these primary laws of motions, then there could not be a motion at all. Eventually absence of law of conservation of momentum will destroy every momentum and motions, and there would be a Nature with no events, no motion, and no Time. This would not create a real world because only motion can initiate further motion. Absence of motion can't create motion. If new fresh motions will be created in such Nature, then it would be a miraculous world which would have no compulsion to obey any laws, and that world would be beyond the scope of science and common sense.

Isn't there any way to imagine such graviton that can bring two solids together by staying and moving between them or connecting them anyhow! Of course it is possible. For that we have to imagine a graviton as a relative absence of momentums. To achieve this relative absence, we have to increase presence in surrounding of the two solid objects. This relative absence if imagined as a corpuscle, it would behave against the newton's third law of motion. It would attract any solid which comes in contact with it toward its side. When it would be in between two solids that relative absence would attract those two solids together. This is nothing but a shadow gravity or push gravity theory, and that relative absence is a shadow formed in a mechanical gravitational environment working with the laws of classical mechanics. So, here is a graviton, and it is a shadow.

To enable the graviton to work within a range of reality or within a framework of classical mechanics, we have to imagine something with infinite numbers of moving 'momentum carrying things" in a space, that forms a Brahm.

If you are accepting the thing Attraction as primary fundamental phenomenon, then you are accepting a miracle in this mechanical environment. Understand the phenomenon of pure attraction as described below.

Real pure attraction:
In this phenomenon two massive existences have tendency to move toward each other without any force pushing them together.

So, real pure attraction is an attraction and not a push if it is a real attraction.
If we consider a space between and around two objects as an absolute blank, then there could not be a mechanism which can push those two objects together. There must be something between or around those objects for any mechanism to get working there. Relativistic spacetime matrix is also something,

more than an absolute blank, because absolute blank is a nothing, and nothing can only have a nothingness and no other property, but spacetime have a property according to GR, so that cannot be considered as nothing, but it may be considered as Brahm. The mechanism of bending the spacetime is not explained in general relativity, neither any description about type of the material is there which is curved or bent by the mass around it. Mass due to its property as a mass, curves the spacetime around it with a gravity and that curved spacetime forms a geodesic responsible for gravity. Here power to create a gravity totally belongs to the mass which gives a vicious nature to gravity which increases in a strength as the mass increases without any higher limit. This vicious nature of gravity creates a room for the formation of blackholes which are structures made by crushed mass with a power of mass itself which mass retains even after crushing itself. This also points to the assumption that a mass preserves its property even after being a singularity. This reminds me a paradox known as Omni-potent paradox. A famous phrase is there "Can a God create a stone so heavy that it cannot lift it?" Here the question should be like 'Can a mass produce so intense gravity by curving geodesics around itself that it can crush the mass in singularity?" If this is possible then it solves a quarry of the origin of the big bang. When singularity becomes so intense after crushing so much amount of mass it loses is property as a mass creating curvature in surrounding spacetime. This would result in abrupt cessation of the gravity force resulting in a blast, a bang. So, actually we can dedicate a power to produce gravity to mass itself which could result in consequences mentioned above but that power of mass forming a gravity itself remains undefined.

Concept of geodesic shows that a curved path in the space-time guides the direction of motion within it, means it directs the moving mass, and it is able to apply force on the moving mass in this way. There must be its structure and structural integrity to provide such resistance and torque, which could not be provided by absolute-nothing. Einstein's field equation just provides a tool to calculate the curvature in a spacetime. So, for the attraction force to be working, one must have to accept that the blank space around two objects can't really be blank but there must be something there. Actually, General relativity suggests Brahm as cause for curvature of the space-time. If we assume the attraction to get working with a field then we have to explain the concept and the nature of the field in detail, and even a field can't exist in an absolute blank without its forming material. Units carrying propagating such field must be defined in a way that it could exist in an absolute blank and there should not be a need for further imagination of attraction force at smaller level otherwise the query remains the same. So, only a presence of something in the space around two masses can explain the attraction phenomena between two masses called Gravity. So, one can't deny Brahm on the basis of the logic and physics. Gravitational force is the proof of existence of material things around and between two masses capable of transferring momentum which can be termed as Brahm. This Point is further explained in the chapter explaining natural forces.

2. Mass integration requires Brahm because only that explains attraction force as a push effect.

3. There is a need of medium for the existence of Energy waves including gravitational waves.

We know the behaviour of the mass in the thermodynamics and in a phenomenon of entropy. When a mass has the energy, mass or its components must have a motion, either linear or circular. Energy can remain only as a motion because "**energy is the motion**" and "a motion is the energy". So, when mass has an energy, it must move. When two or more body of mass with motions if placed together in the absence of any effective attraction force between them, then what would happen there? There is a no need of the law of thermodynamics to understand this. We can think with laws of classical mechanics that when two or more hypothetically self-stable moving mass kept together in a closed blank space, there will be strikes with collisions amongst them. They will be thrown apart. As much the energy will be there, that many strikes will be there. If energy will be enough then the encloser will get blasted and everything will get dispersed in the space in many directions. They all will not attract one another anyhow. So, two or more masses can't remain together until they are being pushed together or there would be an attraction between them. I don't accept the phenomenon of pure attraction because it is miraculous and without a mechanical sense as describe above. It is a mystery and a thing to derive as a result of mechanical process and not a thing to accept without mechanical sense before deriving it as a result of phenomena of the natural mechanical process. Gravity is described as mechanical push in relevant chapter. So, in this scenario there must be a push from the surrounding for the mass to get formed and to stay formed. For the existence of the surrounding force to exists there must be something around the mass. So, space can't be blank space, but it must be full of somethings. Those somethings can either be tiny corpuscles (masses), vortices or energy waves. If we assume them as corpuscles, then again there must be something around each of those corpuscles to make them stable and preventing them from disintegration. So again, surrounding thing must be there, so on and on...up to inner and inner scales there will be a need for the existence of smaller corpuscles to stabilise the higher scaled corpuscles. In second case we assume waves as a source of the force. Now wave is a propagation of the energy in a medium with serial fluctuations. So, medium must be there. We are inquiring fundamentals with this theory, so we will not accept that a wave can exist without a medium, because it is against the definition of the wave itself. Actually, accepting a wave without a medium is a second try to accept mystery as a miracle itself, which is not entertained here. So, in succeeding chapters there is an explanation of energy waves as aethereal waves.

4. Requirement of Brahm is there for interconvertibility of mass and energy.

Medium with a potential energy can explain spontaneous generation of mass from apparent nothing and disappearance of such mass in space or quantum vacuum. If we consider space as an absolute blank with only nothingness, then phenomenon of quantum fluctuations can't explain with conserving laws of conservations.

5. To get the rid of "nothing" we need the existence of something.

As soon as we accept the existence of something, the logic leads us to Brahm. This is explained in the chapter "Honi" a possibility of something which seems like a theory of creation."

6. Galactic centre

In a structure smaller than galaxies like a Quasars, the energy wave fountain from the poles are mysterious phenomena. To explain that phenomena there is need of moving energies in surrounding, and that demands existence of Brahm.

7. Explanation of charge, need of completion of the electric circuit, magnetic field, electric field, gravitational field, …requires Aether or Brahm

Actually, each and everything in existence demands the existence of Brahm but these becomes evident only after proper understanding of this model

2. Honi

Possibility of existence of "something" (Bhav) and its definite certain consequence (Honi).

This chapter is based on a thought experiment on the ground of classical mechanics. This provides an overview of Brahm model. Quantitative calculations can't be derived without the application of math in real situations which require expansion of theory by experimental data incorporated in a theory. Math has an ability to derive quantitative calculations, but math on its own can't justify the concept itself. Math is a tool that needs laws of physics to explain the Nature. Math is an imaginary artificial calculation system which is not structured to match Nature for its operations. Infinite, zero and one are not relatable to the scaled Matrix. One can format the structure of these numbers in relation to Aether-Brahm only after finding quantitative data related to scales. So, without those data in this theory math is not entertained much. If there is a break in logic applied in a concept, math will lead to nonrealistic and miraculous and self-contradictory results with paradoxes due to logical fallacy. So, here logical concept is the mainstay of the theory and math is just applied classically and superficially.

This chapter must not be misunderstood as the theory of the creation or the first and foremost beginning of the universe. This thought experiment is here to find out 1. Minimum requirement that can be enough to create a situation like the present world in logical imagination. 2. Consequence of unstable mass or masses that can provide the insight about the stable state.

This thought experiment explains the ultimate fate of any existence that can be possible. So, here we are starting from the simplest situation and gradually increasing complexity until we may have a situation similar to the present world. This thought experiment also provides a logical conclusion that any possible possibilities of existence must ultimately settle down to a state that is shown in this Brahm-Aether model.

A thought experiment is shown below which can show the possibility of existence of the Nature which can represent the present world at some place and sometime, within it.

There can be many way to reach that situation, but what we want to conclude here is thinking about the minimum requirement to create a situation like the present world.
Let's understand the plot:

Assumption with common-sense for general concept about something and nothing:

We will avoid waves for some time to stay away from confusion.

We know two different type of existence. One is something massive and another is a space or nothing where something massive can have an independent existence and it can have a space for the movement with limited speed only. At the time of assumption, we don't know what this something and nothing are. We also don't know the relation between this something and nothing. We don't know if this something is totally different from the space or is a folded or condensed space or some else modification of space. We don't know if there is something in the space there which is interacting with this massive something without revealing its existence or not. Here there is a gross common-sense based idea about something and nothing. Idea with common sense for something and nothing is just a difference between them. Something has property of mass, inertia, opacity, or translucency with or without the property of refraction. Nothing has only one property that doesn't have a physically revealing property like a mass, it allows free motion of mass within it. It doesn't have a property like gravity, or optical property for refraction until mass create any field within that nothing. So, here nothing is assumed as pure complete vacuum. Two must properties of something includes, 1.) It acquires space, and it expresses itself as mass. 2.) It interacts with other masses.

Let's begin our thought experiment which will reach to reality. We will try to create the existence of Nature including universe from something and nothing with least required complexity.

The existence or Nature can be possible only as three possibilities shown below.

1. There is nothing everywhere, which means there isn't anything anywhere. (only nothing)
2. There is a dense solid everywhere and no existence of Nothing (blank space) at any point. (only something)
3. There are two simultaneous things that exist, one is something and that something is not like option two, so it is carrying nothing within itself somewhere and in the surroundings also. In other words, existence of something which is not uncompressible or absolute dense and alone.

We will inquire each of above three possibilities by comparing it with known facts and analysing it with logic and we will determine if it may fit to form the real world or not.

1. There is nothing everywhere, which means there isn't anything anywhere.

   Is there any need to rule out this possibility? I don't think there is any need to rule out this possibility. A simple argument rules out this possibility.

   "We are, because we exist." If we exist, then we are something and not nothing. So, it is not possible that there is nothing everywhere and there is a no something. I'm something, you are something, and everything is something.

   So, we discard this option as "Impossible possibility" (Asambhav)
2. There is a dense solid everywhere and no existence of Nothing at any point.
   Let's rule out this possibility.

   If the same homogenous dense solid thing is present everywhere and there is no blank space, then it can't be compressible or movable. So, there will not be any motion. So, there will not be any event. Without event and motion there will not be a time. So, the situation will be the same as option one "nothing is everywhere", so it won't allow the possibility of the present world to occur. So, this possibility is also ruled out.

Third possibility is,

3. There are two things that exist, one of them is something and that something is not like option two, so it is carrying nothing within and surrounding itself somewhere.

This is the only possible possibility, which allows an event to occur, that may ultimately settle down to a state which may unclose the real world within itself somewhere, sometime.

Let's explore more possibilities within this possibility. Now we have Nature with something and nothing.

There are three major possibilities here,

(A) Something is surrounded by nothing.

(B) Nothing is surrounded by something.

But we will inquire each of these as one single possibility, because every beginning must have the end, so even if we imagine nothing in something, that something must have boundary at somewhere. So, it will become same as possibility (A). Still nothing inside something has different outcome locally for some time period. Below there is a discussion about possibility (A).

"Something" is a solid and "nothing" is a blank space in a simple definition. That something has boundaries, those are in contact with nothing. If only one something, a stable nonfragile body of mass is there, then there won't be any event to occur. Because even if it would move in nothing, there would be nothing there to compare its motion in relation to it. So, there would not be an event. There would not be a relative motion. And there would not be a present world to be possible in that possibility. But look, normally as soon as there is the birth of a human baby, two other things come into existence, a father and a mother. Similarly, when we assume a solid body, nothing becomes infinite space, that allows a meaningless motion of that only single mass.

This mass has a surface or a boundary. At the beginning we just assume this boundary as a stable boundary, which has mass on one side and a blank space on the other side. But this raises many questions, what forms that solid mass! Is it unistructural or there are units. Are there any further subsidiaries for those units! If yes, what keeps them stuck together? Of course these must be answered, and we will think of it once we have a space for that descriptions.

Now think about the possible properties of that solid.

1. It is a **single unbreakable** unistructural solid. So, it will remain as it is and there won't be anything to be described as an event. There is nothing to compare with it. There would not be any motion, because motion is a relative thing. So, this won't create a situation that can lead to today's world with complexity.

If we assume plenty (infinite) of such solids, moving randomly, they can create events, but there would not be a formation of stable structures without any attraction force if all are of same sizes.

Only If those would be of different sizes, than those could create events and form structures even without any attraction force. Self-stable masses of different sizes may create a Nature that can accommodate the real material world. But here a question has arisen: how different sized solids could be fundamental and self-stable? This more looks like an imaginary nonrealistic situation where the suspicion is there that something is hidden which is responsible for the stability of such different sized corpuscles. Indeed, this situation is the same as Brahm with self-stable fundamental corpuscles and no further query. So, this is a simple form of Brahm described ahead in chapter 20.

Let's proceed to another possibilities.

2. Whether it is **breakable (brittle) or fragile** and there is energy inside it which can lead to its disintegration or produce a **blast** converting it into smaller and smaller masses in the absence of attraction binding forces. (Big -bang?)

Let me explain this possible blast. What we have imagined here is a primary simple mass with no complex properties of the mass like gravity, strong and weak nuclear forces, electromagnetism etc. We want to discover all these other properties of the masses as a result of mechanical processes. So, here mass has only one property that it acquires volume in space. It doesn't share that volume with other masses. This results in a fight for space. When one of the masses tries passively due to its motion to get place of some other mass, then it doesn't become possible without interaction. Now three things are there. Masses, space and motion of masses. Motion is due to momentum or momentum is due to motion, both are simultaneous existence or different terms and describe different aspects of the same thing. What a good thing about momentum is, it is always conserved if the system obeys the laws of classical mechanics. So, what I want to highlight is, if a fragile mass has energy within itself, it simply means it has subunits that have relative momentum and motion in relation to each other. These lead to internal collisions and due to lack of any type of attraction forces, this single mass with such moving subunits will result in a blast according to thermodynamics or common sense.

This will create a situation similar to the big bang. Here it seems that the solid will blast theoretically with a speed equal to infinite because no resistance is there to resist the motion of its fragments. Will those fragments move away from each other with speed equal to infinite? As soon as the motion of its fragments starts, the word infinite speed will not be applicable, because there will be many relative movements, and then there will be a comparison between motions in relation to each other, so any speed will not be considered as an infinite speed then. Even before the blast relative speeds were there and that resulted in blast only. After getting a good overview of this Brahm theory we will come to know why relative speed can't be infinite. But this is just an imaginary situation which is unstable at many points and on many aspects. Still, we are leading with this assumptions to understand a course from instability to stability.

Let's proceed further.

This situation will be like the existence of gas molecules in a space with some temperature and no effective gravity. In this case all fragments will get away from the place of cluster and finally they will move apart and there will not be any formation of mass bigger than gas molecules. That can't lead to the formation of the real world. So, for the formation of bigger masses (clusters of unit masses), we need at least two type of solids, smaller and bigger.

(remember we have not accepted gravity or any attractive force as a primary natural force; so, this blast will not follow a path as shown in big-bang theory)
Now proceed further.

In case of situations like gas with molecules, nothing creative happens due to self-stability of molecules assumed here.

To get a creative situation we have to inquire a possibility where body of masses get fragmented or blasted in smaller and smaller fragments continuously until stopped by any kind of force. Actually, this situation is a natural course because we have assumed a stable mass without any attraction force which is impossible. If there is a no attraction force, then there can't be a mass formation. We are not going to accept any unexplained attraction force. so, for now mass must have to disintegrate in smaller and smaller fragments until any emerging force stops it.

So, here single fragile mass has blasted and fragmented in many small fragments. Those infinite smaller fragments would also blast in a same way. Its smaller components would also blast in its smaller components. This would continue to happen at smaller and smaller scales. This may create a situation like a today's world, but near boundaries mass tends to scatter and soon with the flow of the time all mass would disintegrate in the blank space and there will be scattered smaller solids which will continue to disintegrate with the rate equal to blast. But that won't create a world like today for longer time. Now the existence of that world for a longer time is disputable, because when the event of the blast reach to smaller scale, then much smaller scale, there will be a scale, where a relative rate of time will be very slow. Many years at that scale will be equal to second of higher scale and that would be applicable to all scales. A fraction of time for the big bang will provide a near infinite time at very smaller scale. This describes a situation of expanding universe. Where a single big-bang results in a formation of the world without any fundamental forces.

But what would be there? We can't determine at what scale the blast would be ceased, and the system would achieve non blasting solids or a point masses. We can't understand what would stop that blast! Because if the same process happens at lower scale, then it will continue to infinite levels! And everything will remain blasting for-ever at smaller and smaller scales. Do this create a steady situation even after enormous time? No, we can't find a sure reasons for that. But we find a fraction of time there on some scale to accommodate a world at some level in that process.

A place that can accommodate the real world in this single big bang would be like an expanding universe, where all directional random motions of infinite corpuscles would be there which would create a push gravity and other fundamental forces. It would form structures with stable corpuscles. This would be a universe with an expanding state, and stability of the world would be transient due to expansion that may lead to dispersion of the shadow gravity force and annihilation of all stabilized masses. Here remember that a situation like a big crunch is not included as a consequence of this model because it is not possible with push gravity type attraction force. This model of expanding universe is possible and its potential to create a transient real material world is accepted but still, I want to propose a more stable situation.

Still there is some need to choose a change in possibility for steady state.

If we added gravity as a miraculous attractive force then it could change the situation and help in creating stable masses and direct the process on the track to form real physical world, but we determined at beginning that we will not add any miraculous things or force which cannot be explained in a mechanical sense.

Let's move further.

If we select a finite volume of space within this blasting environment, it will be like a space volume containing multiple blasts at regular intervals, and that volume is expanding in a

Nature due to being a part of single large blast. Another possibility can be imagined if we want a nonexpanding volume. Suppose there are blasts at regular intervals up to infinite distances in all directions and that distances are not increasing, means no expanding universe. This disintegrating and continuously blasting material at smaller and smaller scale disperse in the surrounding and gets mixed in one another's area. Soon there will be a situation when such material becomes moving from all directions to all directions. There will be a mixer of different sized materials which could be grossly categorized into smaller and larger materials. This situation will continue for some time. Gradually bigger masses will disappear by blasting in smaller parts or by strikes with other materials. Biggest solids will only have a maximum size that would allow their journey without strikes with other similar sized materials. Smaller solids will be there because disintegration in smaller fragments is continued. There will be a situation where there is a "gravitational material environment" with moving small and smaller solids with continued and halted disintegration processes upon them. Those will continue moving from every side to every side. Ultimately there will be a situation where the disintegration process on all large solids will stop due to the push of gravitation force generated in this process. More disintegrated bigger masses would contribute to increasing pushing force by generating smaller masses. There will be an equilibrium state where pushing force will become enough that will not allow disintegration of further larger masses. The situation now formed is roughly the same as assumed by Nicholas Fatio de Duillier to create mechanical gravity in his theory. In that situation, there will be a mechanical push which can be described as gravity for the masses. The same push will also stabilize that mass from further blast or disintegration. Try to understand these situation, there was a continuous process of blast of the fragments in smaller fragments because there was no force to stop that process. Smaller fragments were further blast in smaller ones, those are further blast in smaller and smaller. When the nearby fragments blast, their yield products move in directions toward one another's area. But when boundaries of surrounding blasts cross each other, the motion energy of moving fragments would create pressure on the bigger fragments and that would prevent them from further blasting or disintegrating. Such solids will find attraction to other solids due to push gravity. So, the clustering of solids will be there, and it will start a process of the formation of mass clusters. In the process of disintegration into smaller and smaller fragments, there is a push generated in this process that prevents masses from blasting further and various masses remain preserved in this process. We can see here that such push is generated as a natural mechanical process, and it forms forces that behave as attraction forces. These forces form masses and stabilize them. This is not the stage where matter is being formed, this is the stage where components of Aether are still being formed. I don't accept this level at a stage just before the stage where today's world is! I don't accept it at the stage of today's world because it is not able to create a light with characters of so-called EM waves at this stage without Aether and we haven't assumed Aether at beginning to define mass which is disintegrating. Try to form a light like a wave at this stage, and you will realize that that wave will get dispersed in the surroundings. If light is a wave, then it is a non-dispersing stable wave in vacuum. It moves only in one direction and doesn't disperse like sound waves or water waves once it starts its journey in particular direction. So, there is a need for special environment to get such waves. Absence of such wave in a situation at this stage rules out the possibility of today's world at this level if we accept light as a wave. But if we accept light as a corpuscle or structure of mass then this stage may accommodate today's world with variety of mass structures and events.

Before proceeding further from this stage.

Let's enumerate things in existence at this stage.

1. There is an infinite space.
2. There are infinite number of blasts at some intervals in that space.
3. Push gravity is generated in this process.
4. Two types of materials are present, one of them is disintegrating or blasting material which is at smaller scales which is still not stabilized by effective push gravity at that

level because only smaller particles can provide push gravity to larger ones. The smallest would continue to be blasted into much smaller types... when smaller would be available, it would stabilize some of them at that scale. The same thing would happen to that smaller masses also. So, at every levels most solids would get blasted and few of them would survive as stabilized solids. Such stabilized solids form the mass portion of the material. Those will play key roles in the formation of Aether and vortices in Aether later on.

These were the things created by multiple blasts of material things in a space. Let's choose a place in this space and describe its environment.

There were infinite blasts in space in all directions at different distances from this place. After some time, there is a situation as above where some masses get stabilized and achieve non blasting state due to stabilizing pressure generated by push gravity phenomena created by smaller moving masses. That pressure will stabilize these bigger masses. Such masses will get attracted to one another and will form a cluster that may result in a structure "hollow". This structure will form a mass cluster. This structure would grow to optimum size by getting further mass from surrounding powered by push gravity. Now there could not be a further increase in the size of such hollows, because gravity push is not enough to add more mass to the hollow because clustered masses also have a kinetic energy, that will resist push gravity. So, these hollows will stay at distances from each other and will not get fused due to gravity. Let me introduce this structure hollow. You can compare it with stars of the present universe. Stabilized solids are clustered by push gravity. They form a globular cluster. Gravity push continues to add further stabilized solids to it. That stabilized solids have a kinetic energy due to motions. After adding enough stabilized solids, gravity push will have an equilibrium with kinetic energy of stabilized solids of hollow. That will not allow further addition of material to that hollow. Globular cluster will achieve a stable maximum sized state and now it is called hollow. During this formation process, hollow would get some spin due to unequal distribution of kinetic energy among cluster forming stabilized solids. This spin might create an orange like shape distortion of hollow with axis formation. Understand the push gravity, only fraction of the moving fragments produces the push gravity and majority of smaller fine fragments just pass through this material without interacting with it. The flow of smaller solids can be termed as flux. The portion of flux which don't interact, don't form the gravity push. Only fraction of flux produces a gravity push. This push is called effective gravity push and in short gravity push. Hollows are formed by many smaller scaled hollows and tori. Torus is a structure formed by arrangement of solids in a ring like or disc-like shape which forms a magnetic field around it. Torus formation is a separate topic in a chapter 5. Due to spin, the magnetic field of tori parallel to spin axis get united while magnetic fields. In other directions they don't unite. This magnetic field is curved or bent energy flux due to gravity. Now spinning hollow is a structure with energy field surrounding it and is now a polar object with axis and poles. Positive and negative poles are due to relative direction of the spin that contributes to the twist of the formed field. If Stabilized solids are clustered by gravity push and don't form spin enough to create Torus, then it forms Hollow. Now when we think of clustering of polar spinning hollows having fields and poles, we will get a vortex which is a galaxy-like structure. This vortex will form a torus structure. Such tori when arranged in a lattice it would form a material within a big globe. That big globe would be like a primary corpuscle of the higher scale. It will be stable structure and become the unit of the Matrix. This material formed by such globes would be like an elastic medium with resilience. It will have energy fluctuations from all sides and motion of micro solids from all sides. This is a material which should be called Aether. Because at this stage, this material has units and so it will behave like a Matrix with elasticity and resilience. There can be waves in it, there can be wave-clusters within it. There can be a formation of vortex of wave-clusters also. When wave-clusters and vortex of wave-clusters will get formed in this Aether, then it would behave like a resistance for moving waves and flux materials in that matrix. In the formation of stable environment of this Aether units, when two large tori will get attracted to each other, it may fuse with each other. If impact is strong enough, it will result in bigger torus. If it has resulted in larger torus, the gravitational push may not be enough to stabilize that bigger torus, then extra part of it will get its way in surrounding and only part that can be stabilized by gravitational push will form the torus of possible maximum size in relation to available strength of the gravitational push and material. Such strikes, fusions, reformation will get repeated until there would be a peace, and calm state will get achieved. When calm state will be achieved, tori will get attracted to each other but that won't fuse to each other because they lost much of relative linear kinetic motions. Tori have poles, field of energy, so here attraction will not be only due to push gravity, but field interaction would be there, and you can compare this field with magnetic field and force with magnetic force. So here attraction will have two components, push gravity and magnetic force (energy field interaction). This process will get repeated again and again for enormous time. And there will be a situation when there is a steady state where such tori units are placed in a pack situation with touching of their field areas, and it will form a continuous material that would form a stable hollow with outer surface, crust boundary and inner core. And such hollows would form the unit of Aether again at higher scale. This Aether is liquid like or elastic solid like still compressible, because field would allow relative movements, if units will be acted upon by force, relative motions of those units of material will be there. The field will also be compressible. There would be resilience which allows waves to be formed. There would be a complex-wave cluster formation which would behave like primary masses and there would be a vortex formation and further torus formation with these masses, this results in hollo formation and torus formation at higher scale. So, though we have started with undefined mass and blank space, we could now figure out that those masses were formed by wave clusters within fine scaled finer Aether. Thus, Layers and layers of Aether forms scales. Layer is termed as a

"Ther" and "Ae" Means "that" in my mother tongue, So, if Aether would have origin in my native language It would mean "Bunch of layers". When plenty of energy moves to all sides from all sides, there are continuous fluctuations in Aether, and that vibrant fluid is now called Brahm. There is a separate chapter describing this Brahm.

So, there is a need for an infinite number of solids to be there and in a motion for the creation of the situation of the non-expanding present world and that will provide a long-lasting steady state. Only one blast is sufficient to create a situation of the expanding present world, but it would provide a transient stability only and not a steady state.

This Aether world can have an aethereal waves, it can have atoms and matter also. Material cluster, Hollow and the Torus as a whirlpool or vortex of wave clusters in this Aether behaves as a primary mass units with magnetic components. With compound gravity these units can form a subatomic particle and, cluster of these subatomic particles can lead to formation of nucleus. This nucleus is the atom, when it includes its field and its rotational kinetics with its interaction with surrounding energy flux, it forms complete atom.

In respective chapters concepts for **electricity** is described in relation to the spin energy of nucleons and nucleus of atoms.

Aether is not imagined as a homogenous structure for infinite distance. Saccule formation occurs in a same way as hollow formation. Such a big, large sac that uncloses plenty of tori or galaxy-like structures is called universe vesicle or brahmand. It is the nature of push gravity that it arranges clusters in sacs and not continuously due to its limited strength and shadow effects.

Now this Aether is a medium creating a closed system that holds optimum energy at every unit of it. This closed Nobel system is **"<u>Brahmand</u>"** a non-empty hollo structure. In this Nobel system, there is a balance between medium and structures within systems. Normally both of these are in an equilibrium for energy. When any event tries further to raise the energy at any point, that point and surrounding will resist that act and will provide feedback or back thrust. Thus, this energy equilibrium is maintained with some buffer margins. Such medium also behaves like an elastic medium with resilience. Now this is the ideal situation for the formation of aethereal waves (EM waves and gravity waves). Such waves are compound waves which have oscillations of aethereal units in both transverse and longitudinal directions. At a very long distance such waves suffer a transformation due to the nature of Brahm and get constricted. This phenomenon is explained in a chapter 6 in a topic "**Wave transformation phenomenon**".

It means light waves are not considered as immortal or free of ageing effect here. Gradual transformation is there, with light also, when it travels in Aether. Not just materials scattered in the path of light affect it, but Aether itself applies changes to the morphological structure of light waves.

Let's compile the attributes we have found during the above thought experiment.

1. A stable mass needs a push gravity around it, otherwise it disintegrates into smaller fragments. During this process it may radiate waves due to kinetic energy transfer to surroundings.

    So, mass is a compound cluster of polar and non-polar structures formed due to cluster formation of survived material at each scales, stabilized by push gravity.

2. There cannot be a non-dispersing wave like a light without a special environment which can be provided by Brahm like medium only. So, formation of light wave is not possible until the medium with resilience is there. Environment of only moving particles without a medium with resilience can produce clusters but may not the light as a wave. But light can be possible here as a fish projectile which could be a particle form of light or can be a particle with field also. So, Aether is indeed for light as a wave, but not necessary for light as a particle form.

3. There is no need to assume a mysterious attraction force like a strong force, weak force, or fundamental gravity, all that can be an act of push gravity at different mass density and

distances. A simple world with mass and emptiness is enough to explain phenomena that can replace the requirement of such unexplained multiple mysterious forces.

4. Star or galaxy-like structures can be formed in a medium with resilience due to the effect of push gravity.

5. In absence of push gravity, a single mass cannot have a stable surface direct in contact with the absolute space. If we accept such point masses as self-stabilized masses, it will form the lowermost scale of this model, and we don't have any problem with them. Such masses will stabilize higher masses and thus it will provide a gravitational environment. But this assumption is a result of our inability to inquire about the real situation. An essential assumed property of mass to disintegrate in fragments is not sustained in such assumption, and I think this is not essential because we can get enough survivors among disintegrating masses at each scales powered by push gravity.

6. The push gravity has a finite intensity when it gets applied to the mass having specific density distributions. When the amount of the mass with same density distributions exceeds certain limit, effective or interactive flux forming push gravity will become unable to hold it together and penetrate beyond some limits. Non-interactable flux will be there, which will travel through all material at that scale. When mass units have a kinetic energy and nature to repel each other due to their oscillations, spins, and other motions, after certain limit, then it will not be possible to add more mass to that cluster. This will determine the maximum size of the mass cluster. Density factor will also work here. Distribution of actual, obliterating vortices in a mass will decide how strong the push will be between them. The flowing energy carrying units which will be fine enough that a mass couldn't obliterate its path, will not contribute to the push gravity at all because it will be there, but there will not be effective interactions with porous mass. So, the limitation of push gravity will depend on the property or density of mass also and not only on the amount of energy flow alone. There must be an unused fine energy flow at all levels of mass formations.

7. Polar mass units can be formed in a model described above which have a poles where energy waves continue to pass and create a field. This can affect the orientation of such structure when placed near to one another. This can be a base for magnetic field and basic magnet units. This can also provide logic for the instability of the nucleus of the atom when there is an imbalance between the number of protons and neutrons.

If we accept this model, then we have to reject following things,

1. We have to reject the existence of Strings or solids with smooth stable surfaces as basic units with shapes like a filament. We can't imagine a string or solid with smooth stable surface-like structures as basic units because there must be a question "why it can't be disintegrating in a subunits?" Imagining absolute mass and absolute blank space is not possible within the scope of this Brahm theory. If Strings has no subunits, then what gives them that shape? If they are the fundamental basic units of the mass, then what causes them to form clusters and structures! If vibration of such strings provides them with characteristics, then there can be millions of types of variations that should result in millions of different types of particles, but we don't find these many varieties of particles. Here there is a clearcut difference assumed between mass and space. No energy relation between mass and space. How could such mass get attracted to other such masses if there are no smaller moving units forming gravitational environment. Are all strings of identical sizes or of different sizes? If such two types of masses are there one smaller and one bigger than what differentiates them from each other and what determines their sizes! So, when we try to simplify the thing with strings, it actually becomes complicated. So, imagination of such masses with smooth surfaces in any shape doesn't provide a model of the physics with satisfactory explanations and clarifications at fundamental level. To get an easy way of formation of particles by its vibrations seems good but not a plausible idea. Interaction of vibrating strings would ruin them all. When two strings come in touch with each

other without fields, their vibrating properties must get changed. Imagination of vibrating Brans in M theory can be compared to tori or galaxies. But in this Brahm theory that tori are not fundamental units and there is a description of tori in detail.

2. If graviton is responsible for gravity, and gluon for the strong force! And, if gluon stabilizes atomic nucleus, then what stabilizes the gluon itself and subatomic particles themselves. What is the force that doesn't allow annihilation of the neutron or proton in the form of energy, what holds integrity of such structures! If the answer is gluon, it is not possible as explained ahead. Attraction can be produced only as push gravity mechanism mechanically. So, it is unimaginable how a Gluon can hold subatomic particles together and how quarks could be held together. Neither logic nor math can have solutions for this. So, when one doesn't understand the phenomenon, one just adds a particle with characters that can do that work mysteriously, then it doesn't formulate a science theory, but it just makes a fairy tale. Mechanism must be explained clearly and with realistic concepts to get accepted by a logical realistic thinker. The only way to imagine such gluon that can produce attraction by interaction between subatomic particles is imagining them as a shadow in a push gravity. Only that can behave like a gluon or graviton. But then it will not remain different from Aether-Brahm theory.

So, with the common sense acquired from this model we will reject a single existing proton or neutron in an absolute space without surrounding compressing energy flux. There must be a compressing energy flux around it for its stability. (energy flux= what can produce a push)

So, we can't accept a gluon as a self-stabilized structure. Similarly, we can't accept graviton as a self-stabilized structure. We can't accept strings or Brans as basic most fundamental units when it doesn't derive gravity and other forces as a mechanical phenomenon.

3. Non transforming waves.

As it is explained ahead in the topic energy waves in a chapter 6, Each and every wave must have decay or transformation. So, Light can't be an exception. The base for expanding universe a redshift is not an actual redshift but is an effect of the wave transformation or wave compression phenomena. This effect rejects the expanding universe and a single big bang theory. This also challenges the present map and estimated size of the structures in the observable universe. This theory doesn't say anything about the maximum age of the universe, but it says that minimum age is far more than estimated age in a standard model. Limitations of the light waves don't limit the extent of the universe. Limited size of the observable universe is due to limited life and limited visible form of light during its course in space.

4.   Inertia as internal fundamental property of mass:

This theory rejects inertia as an internal property of the matter, but this theory shows that inertia is the equilibrium change between mass and surrounding energy flux. This theory shows relation between newton's first and third law of the motion as described in the chapter "Inertia."

5.   Star and Blackhole with no higher limit of mass.

The mass is the result of cluster formation due to compound gravity including mechanical push gravity component. There is a finite intensity and concentration of energy flow producing mechanical push. so that can't provide infinite pressure that can make cluster of infinite amount of mass! at a fixed density level of the mass.

So, if the star gets an amount of mass which can't be held by gravity, then it will lose mass as extra radiation and flares or get blasted. So, at the particular place, according to the strength of the energy flux producing push effect, maximum mass of star can be determined accordingly. The size of stars which are calculated previously with the assumed distance based on redshift, should be recorrected using logic of this theory, then average size of the stars should be recalculated. Centre of the galaxy is described in the chapter "Structures in Brahm", So, that centre of the galaxies are the places having less and asymmetric gravity instead of highest gravity. You can also predict plenty of predictions using common sense generated with this model.

So,

With the insight availed from this Brahm-Aether theory, it can be predicted that what can exist and what can't, with strong logic and arguments.

So, any possibility has a same ultimate settlement in a gravitational environment as clusters, energy fluxes, energy field, push gravity, magnetic forces and lattice formation, that ultimately lead to Aether medium with materialistic expressions. Undefined mass which is shown here as a non-blasted preserved material solids have a fate at the centre of galaxy which ultimately clears it out at that level. Thus, it recycles material. This have explained in the topic galactic cycle. That preserved material is named "Bhut" in this theory. What that Bhut could be is explained in relevant chapter ahead, so that undefined thing from what we have started will no longer remans undefined and we can find it as a wave cluster within Aether, and the thought experiment which mimics creation actually acquires a part of the cycle only. That was the reason I declined that thought experiment to be considered as a theory of creation from real prime beginning. This theory supports the omni existence existing forever that harbours all, and don't seek for the first ever creation or formation of most fundamental thing.

Note: *Though there may be different models, which may provide explanation about the existence, but the natural and realistic one is explained here. There may be a more complex and more developed state of the existence there in reality and that may be providing a simulation mimicking earlier natural state. We can't rule out this possibility at all. We can't rule out the possibility of unexplained existences like Soul or God, because non interacting flux may have potential to create a finer structure not or poorly interacting with this material world, and that opens a window for the existence of supernatural things and extra dimensions also. but all these seems like a fantasy, so such things are not included in the explanation of this theory where only primary simple and natural model is explained.*

92

3. Revisiting Brahm.

In the previous chapters it is explained why the existence of Brahm seems essential. It is explained in this chapter that the reasons which make Brahm essential to exist also point outs to its properties.

If we accept the concept of push gravity about gravitational force, it suggests that energy in Brahm is continuously being carried and propagated from all directions to all directions. Brahm is not a thin quiet medium as imagined like a historical luminiferous aether, but that continuously transports energy in an abundance. So-much energy, that could hold matter of massive stars together, that could hold matter in galaxies together. Which can form a dense atom with highly energetic protons in the nucleus. The stability and integrity of all these structures are possible only by the push generated by the energy flow present in Brahm. That great amount of energy resides at every point in Brahm, still it is not sufficient to create chaos, that just creates orders. Matter is also a complexly spinning state of the waves within Aether itself, that also represent energy within Aether.

In the succeeding chapter we will also see that primary mass is also not a separate entity, but a vortex of the waves in Aether medium. Thus, it forms a barrier for other waves and flux. So, this characteristics of the vortex of wave clusters in Aether is called mass or matter. In fact, it is not such simple like a loop current in Aether, but it is a complex cluster of vortices made up of cluster of waves.

In the succeeding chapter in which structures in Aether are described, there is a description of the structure called "brahmand". This structure Brahmand is like a body of the mass, but it has an outer soft boundary, it has a thick area called a crust and a core. This brahmand is an aethereal structure, and there is Aether inside it, with it and outside it. The energy being transmitted in Brahm is in contact with this Brahmand and actually that energy stabilizes this object. So, Aether is a medium having energy as a wave and it is saturated with the energy in such a way that it provides a Nobel system with moving energy flux in such abundance that they don't absorb net energy from the matter and provides stability to the matter. Brahm is so saturated with energy at every point that, when deflection is initiated with desired frequency and power, then it forms a wave which moves in a single direction and doesn't disperse or spread in surroundings if deflection force is within a range.

This feature provides base for the electromagnetic field imagined in modern physics to explain light as EM waves. Environment outside Brahmand and inside Brahmand must have vast differences. When we think of environment on the earth, outside earth's environment in solar system and environment inside heliosphere of sun when compared to environment in interstellar space beyond heliopause they must be different due to field and shield effect provided by structures. We don't know about the Nature of gravity at inter galactic space at all. So, we can imagine that environment at space in inter Brahmand space may be harsher due to scale changes. Means wall of Brahmand prevents course corpuscles to disrupt orders within Brahmand. It behaves as a shield or mesh that allows fine flux but not a course one.

The waves in Aether have a frequency, width, and a wavelength. For example, let's see the light as a wave. Till date light has not been understood completely. But still science knows some pure facts about light. It is the fastest known moving thing around us. It doesn't disperse in a space, like a water wave gets dispersed in a water or a sound wave gets dispersed in the air. Because of this non-dispersing property, it doesn't lose its energy soon and thus it can travel a long distance for Billions of years. It is clear that an Aether is such a medium that doesn't take away energy from the light wave at least for a very long time or a very long distance. With a telescope one can see or create an image of galaxies billions of light years away. This supports the assumption that all these things may be in a closed but a Nobel system brahmand. Brahm in our world is Brahm within Brahmand, saturated with energies and don't absorb energies from light wave. Still light has an interaction with Brahm and this phenomenon is called tapering or transformation of waves. This phenomenon is described in a chapter 6 and 9. Thus, Brahm has a property to interact with waves such as light.

Few briefs...

Aether is a medium forming that carries everything else as waves within it.

Mass is a vortex of waves in Aether. Energy is aethereal waves or momentum of moving masses.

Aether propagates light. Moving masses and waves including Aether form Brahm.

Energy flow in Aether interacts with the mass as tide of momentum in Brahm which get reflected as an inertia. Mass is complex of waves, and it also have a momentum. Balance of the pushing flux including tides to the matter decides its motion within a Brahm in relation to Aether.

Aether has an interaction with energy waves like light waves, gamma waves, other waves categorized as EM waves in modern physics.

Mass can be generated from energy in Brahm-Aether, and mass can get disperse as an energy within Brahm because it is nothing but a cluster of smaller vortices of waves in Aether. In a supernova blast, energy radiates in surrounding. If extra energy breaks equilibrium between pushing force and internal kinetic energy within mass cluster, then it may become unstable and blast in fragments, during this process radiation energy may get annihilated as a turbulence in Aether, leaving small fragments and freshly generated aethereal waves.

There are structures in Aether which behaves like aethereal units and distributed in a hexagonal close pack symmetry. This may be the reason behind the Nature of the energy waves as spring shaped helical waves. Energy waves are spring shaped helical waves and thus they have two types of polarity. Clockwise in the direction of its motion and anticlockwise in the direction of its motion. Such polarity supports these assumptions of Aether's property. Aethereal units are arranged with compound gravity including push gravity and magnetic component. Thus, they have an entangled positions with each other forming elasticity and resilience and a deflection of one causes a series of events.

4. Gravity, Other fundamental forces and electricity

Gravity

Gravitational force

Definition: when two or more resistance like couple waves or cluster of waves in the Aether are there in a gravitational environment made by moving flux in all directions, they obliterate a path of flux moving through them this results in a common shadow formation in a field of flux that makes a pressure gradient and results in a push for resistances toward each other or one another. This pushing effect that tries to gather resistances in Aether is called Push gravity or gravitational pushing force.

Along with torus formation these Gravitational push turns into magnetic force which along with gravity push forms masses and matter.

Gravity with magnetic force and drag due to reflected flux together forms a compound attraction force.

General belief: Gravity is believed to be a force which causes an attraction between two masses. Sir Isaac newton gave the law to calculate the strength of this force. Albert einstein described it as a curvature of the space time and its field can be calculated with einstein's field equation in general relativity. It is the force that bounds masses of the matter together and that allows an earth to orbit the sun. It forms galaxies and stars. So, gravity is a force we observe and feel in routine day to day life.

Compound attraction

Here,

We would like to explain gravity as the apparent attraction force between two or more masses.

There are three components of this apparent attraction force between two masses.

1. Mechanical Push Gravity.
2. Magnetic force. (energy field interaction)
3. Drag.

Let's describe it briefly.

1. Mechanical Push Gravity component.

Atoms of the matter has a hollow-porous structure, so, gravity push is very weak. Nucleus of the atom has a mass, relatively condensed mass in comparison to cluster of atoms or molecules. This means the majority of the mass within cluster of molecules is distributed in a relatively very small area called nucleus and major portion of the volume acquired by the atom is devoid of mass. So here in a nucleus there would be much resistance to the flowing energy per square unit of volume.

So, push pressure is much higher at the nucleus among subatomic nuclear particles due to denser mass in the nucleus at very near distance. This push works same as a push gravity if we neglect a magnetic component of the strong force.

For example,

Atom to nucleus volume ratio is 10^{15}. There is a huge difference in gravity between two nucleons or nuclear subatomic particles like proton or neutrons in the nucleus and between two atoms. Actually, I am not talking about the attraction force acting at nucleus or subatomic level but trying to explain the weak nature of an apparent attraction force between two atoms of the element which is categorized generally as a gravitational force. Here I want to say that part of the strong force and gravity is the same thing but due to difference in the density of atom to nucleus or subatomic particle it has difference in strength. Strong force is stronger also because magnetic component is stronger at small distances.

So, this mechanical gravity push component plays a major role in the formation of the compound attraction. Let's understand it properly.

Push gravity mechanism

Here Gravity is explained similarly like a Kinetic theory or a shadow gravity theory. In this explanation Gravity is not considered as a fundamental force instead it is described as a push or thrust from surrounding environment that pushes two masses or resistances in Aether toward each other.

Brahm is derived in this theory as a magneto-gravitational environment with Aether. Which have moving energy flow (flux) from all side to all side. This flux may be due to aethereal waves or corpuscles with momentum. I would like to use the term "flux" to describe this moving momentum carrying, pushing things in general. This creates an environment that creates a mechanical push gravity and magnetic effect. If we consider only gravitational part, this theory is nearly same as the mechanical push gravity theory of Nicholas Fatio. The only difference here is, energy flux which could be made up of moving small masses, small vortices or aethereal waves instead of calling it a flow of hypothetical inframundane particles. Further advancement of this theory is explaining torus formation, magnetic field, electricity, strong force, weak force and effect of shadow gravity at a centre of galaxy and centre of the quasars. So, this is not a theory that just include Fatio's push gravity. As per suggestion by Kelvin, this theory also included an aethereal vortex concept but not as a primary simple aethereal current loop, instead, this theory shows a scaled compound structure of vortices of aethereal waves as a primary mass. So, push force in this theory not just explains gravity only, but it explains other forces as well. It also provides a mechanism of formation of mass and matter and actually everything. Let me explain primary mechanism as below.

If we imagine one body of the mass steady in relation to Aether in a Brahm at one place, that one body of the mass will not have any effective movement relative to aether due to equal push from all directions over all its units. It will not move to any side relative to observer steady in relation to Aether if push from all sides would be equal. There would be impacts of energy flow or flux from all sides and it would create a push or thrust on each resisting units of that mass. But push from all side will cancel each other out and there will not be any net deviation toward any direction.

But imagine two such bodies of the mass in a Brahm. Suppose two are placed side by side as shown in figure 4.2. The left one is named as "A" and the right one as "B". It can be understood from porous structure of mass that "A" would obstruct only very little portion of the flux from all sides and there would be interaction and transformations of that flux. Units of this flux would get reflected back in surroundings or would get some directional changes, but it would not pass completely without any interaction. Understand this thing, here I am not saying that mass absorbs momentum energy. Every aethereal wave couples in a mass are bound together due to force from the surrounding push. Surrounding push thus binds unit masses, all vortices, all wave couples. Energy within those spinning things continuously counter acts the pushing flux. Thus, pushing flux continuously acts upon matter units. Continuous flux provides a continuous push. There would be an obstruction to the linear flow of the energy flux here by putting two mass side by side. There would be less pushing flux on the opposing part of A and B. Here B is obliterating pushing flux coming from the right side and the similar thing is also be done by A. The area between A and B will have less pressure of flux, and other sides will have comparatively more pressure by flux. The pressure gradient is being created here. This will create a push on A and B toward each other. Thus, they will be moved toward each-other, if their motion will not be restricted by any other force. So, this push effect between two masses created by surrounding environment is the explanation of mechanical push gravity.

This is the mechanical explanation of the one component of the compound attraction force between two masses.

It should be taken into consideration that matter is a hollow-porous object with much of its area having no resistance to pushing flux. Not all energy flowing in Brahm is creating gravity, only the amount of energy that is obliterated by the matter is forming gravitational push. This means there is much more energy flow in Brahm than what is reflected as a gravitational push, and magnetic effect supports this concept. It depends on the nature of the energy carrier that if it

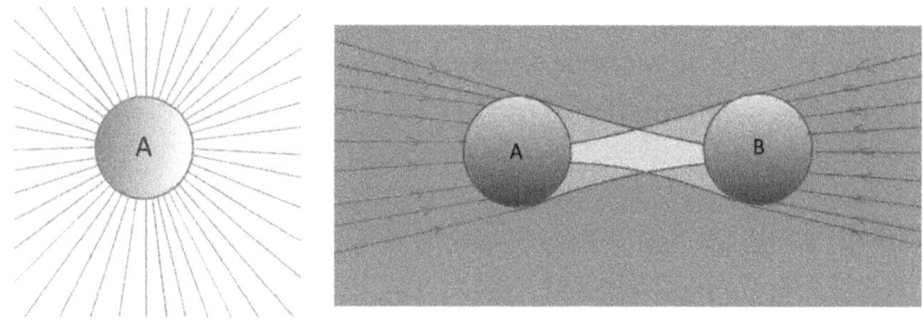

Figure 4.1   Figure 4.2

would interact with matter or not. Matter is a cluster of dynamic components distributed in 3d space with so much space between them. That's why push generated as mechanical gravity is not just at the surface of the matter, but at every resisting part of the matter. Push that stabilizes the mass structure is much more than the push forming push gravity, that's why shadow of such gravity push doesn't disrupt the matter structure in normal circumstances, and pushing flux remains able to penetrate very large cluster of masses.

Please note that here we have shown that gravity is a push from outside. So, its strength depends on the energy carrier's flows (flux) in surrounding and through the masses. It also depends on the concentration of resisting units of the mass. Denser the mass, more will be the push. So, the more mass within unit space, the more push is the rule. Farther the mass, less is the push, because of conical shadow of obliterated energy flow. It has a higher maximum limit of strength. As we know a normal matter is made up of atoms and atoms have very less mass compared to its size. It is empty (no mass) at most of its area where nothing is there which can be categorized as a solid mass. So, matter is a 3d mesh like material which has points where mass exists otherwise it is made up of empty space having no mass. Distribution of mass in this mesh is indeed not equal.

When mechanical push gravity works on such matter, its effective strength which is just a small fraction of its total strength produces only little push compared to the real potential of the total energy flow. The other factor that limits the effective strength of gravity is distance. Because atoms of the matter have a dense mass in nucleus only, when two atoms came in contact by push gravity, two nucleuses stay far away from each other due to size of atoms. So, effective gravitational push between two nucleuses becomes very less compared to similar push between two subatomic particles inside the nucleus. The difference in the strength of push gravity is so huge that push gravity between two nucleuses (atoms) is very weaker compared to push gravity between two subatomic particles within the nucleus. This is an explanation about gravity as a component of strong force in the nucleus of the atom. Gravity is always there between two masses as a primary phenomenon of push produced by flowing energy flux. But sometimes other forces play a major role that apparently masks the gravitational push or override it. Shape and distribution of the resisting units in the mass also plays roles. Suppose there are two balls put together in a space by touching each other and in other case plates are formed from these two balls and placed together facing flat surfaces. Gravity will be more in case of two plates, because resisting units of mass are placed overall nearer to one another in case of plates made possible by their shapes. So, shapes also do matter, and we may calculate gravity strength just by assuming the centre of gravity roughly. Distribution of mass matters. There is a finite amount of energy radiating in Brahm per square unit area of Brahm. Strength of the gravity will depend on that amount of the energy. Mass and molecules have kinetic and heat energy. If we continue to add mass to one body of mass, then at some point the pushing energy would not create a push enough to hold that much mass of the matter, it will allow the mass to get escaped. This is because the push gravity decreases gradually to get penetrated inside that mass. During this process direct pressure of the mass units like atoms would be increased and that may alter the type of the matter. This would create a most dense area called accretion layer. The inner part of that body of the mass beyond that accretion layer will have less restricted kinetic and heat energy and it will provide pressure from inside of that body of the mass. When such body of the mass exceeds some limit of the mass, it will explode or start losing mass. If addition is gradual, that will create a balance with the flux, and that mass will be stabilized in size at maximum mass point. This is not a simple thing as it seems. Actually, there will be many transformations of mass in this case what will ultimately yield this result. Let's think. Suppose there is a big cluster of mass made up of elemental matter. On the surface of such a cluster, atoms have unequal stabilizing forces. Near the outer part of the cluster the atom would be under more push or pressure, while at the inner part of that cluster beyond accretion layer, atoms will lack that direct pressure. But in that cluster atoms would be pushed against other atoms beneath them. That would produce a fusion reaction at some depth in case of enough mass, and it would free energy that would result in radiation. Such fused nucleus would be heavier than other atoms. It would find its path toward the inner side. At the site which may be the centre of that mass cluster or a place nearer to centre of that cluster, again there would be an area of less gravity. That would allow annihilation of fused nucleus into subcomponents, which will move toward surface due to gradient. So, this type of cycle will be there. which ultimately limits the maximum size of the cluster. If sudden addition of mass would occur to this mass, it would disrupt or blast the mass in smaller fragments with energy radiation. This can be considered as **"radiation force of the mass"** which allows supernova to explode, or which limits the size of the proton or neutron and not allowing addition of the further mass to nucleons, proton, neutron, and stars. This force is a manifestation of the limitation of mechanical push gravity. Some other factors also play role in the instability of certain mass clusters and that factors are described in sections of Strong nuclear force and weak nuclear force. The maximum possible size of mass cluster like proton, neutron, nucleon or unit masses etcetera, provides the base for the quantum nature (specific sizes) of particles and energy quanta. Such fixed sized units at smaller scale are responsible for energy quanta related with certain process or phenomena.

So, here mechanical pushing force between two atoms or between two body of matter is considered as a push gravity component.

Mechanical push gravity between two subatomic particles in a same nucleus can be considered as gravitational component of the **strong nuclear force.**

In case of Strong force and Gravity, one seems strong and the other seems weaker. But both are the manifestation of the similar phenomena of mechanical push gravity. This is an explanation about gravity as a part of strong force in the nucleus of the atom. But strong force also has magnetic force component similar to magnetic component of compound attraction force described in a topic strong force.

2. Magnetic force component. (energy field interaction)

This component is very weak in case of force between two matter body or two atoms, because the fields of atoms are weaker in unexcited state. Now this uniting energy field interaction is due to bipolar form of the massive structure called torus. An atom is not a torus, but its nucleus is a cluster of such tori, and it forms an atom, that has a compound magnetic field around it resulting in no net magnetic polarity generally. This interaction is between energy fields. So, this should be termed as a magnetic force. It behaves as a proper magnetic force in the case of some atoms which have nuclei arrangement forming a proper magnet. Otherwise, this force act as repealing force only, and that is one of the components, that don't allow the fusion of two atoms or two nuclei. Means this is an interaction between the fields of two atoms, not allowing its nuclei to come in contact. The spin of the nucleons that forms torus field in Brahm, also helps in providing shield to the atom that prevents fusion to other atoms. We recognize this shield as an electron field. This component has more importance when we are talking about nearby atoms and has little or no importance when we are talking about two matter body far away from each other in case of non-magnets.

3. Drag component.

Brahm has all directional flows of energy or momentum carrying things. If there is an imbalance somewhere it may cause a drag. When such flux units strike with spinning mass, reflected flux units acquire angular momentum and twisted path, that may produce a drag effect to the mass moving in such field area. For example, mercury around the sun. This component is not considered as a routine part of the compound attraction force, but it has an importance in the case of thinking about gravitational field around the huge spinning masses.

Do, Mass drags Aether around it due to gravity? My answer is "I don't know", what dragged around spinning mass is a reflected energy flux of Brahm is sure thing but if there is any drag for the units of Aether itself or not is a question, and if it is there, then up to what extent is a question. But Aether has an elastic property and its units are compressible. Unequal pushing flux concentration around the mass due to shadow effect may cause distortion in the shape of Aether units. We don't know the nature of the reflected flux by mass. In case of spinning heavy mass, such flux would affect the shape of Aether units in the direction of its reflection. This may contribute to time dilation in that area and also to the drag on mass moving in that area. Mass as a vortex formed by whirling aethereal units in a vortex is also a possibility including Liquid state aether. But I have described substance Aether with no internal currents of units of Aether. This helps to achieve mass as a wave and allows time dilation for moving mass in relation to Aether. If This is possible in a liquid like state also then there is a no objection for the Liquid like state of Aether. But the incompressibility of the Liquid can't be incorporated in such liquid like state of Aether. If liquid Aether is there then there could be an aethereal fluid current also which can drag aethereal structures and waves within its flow. But I have not included this possibility in major descriptions and accepted Substance Aether in which a mass is a complex aethereal wave-cluster and not complex loops in fluid aether.

So, this was the description of the compound attraction, with major component as a mechanical push gravity in case of far non magnetized objects.

Other fundamental forces

Like Gravity (mechanical push), Strong force is also explained. Due to more dense mass at subatomic level in a nucleus due to less hollowness, mechanical push gravity is stronger there at those little distance, which contributes to strong force. From the positive charge of the protons, we can imagine that Protons have a physical spinning property that creates a torus like field around it. That field causes field interaction between protons. Neutrons help to decrease repealing interaction between protons. There is a requirement of neutron in the formation of the stable nuclei of atom. When limit of further addition of protons and neutrons is reached or unstable combination is there, the pushing flux in Matrix will not provide enough push to keep them together. So subatomic structures are thrown away as components and there is energy wave production during this fission. This is an explanation of the weak force. Weak force is nothing but, either a limitation of the push gravity or an interaction of fields and motions resulting in repealing force due to unstable combination of subatomic particles.

These forces are explained below.

Strong force

Definition: When matter with high density is placed in a cluster called nucleus with the push gravity and magnetic force, these forces together termed as a strong nuclear force or a strong force.

Strong nuclear force.

In a standard model of the physics strong nuclear force is assumed as an attraction force that binds quarks together to form bosons including neutrons and protons. It also works as an attraction force between proton-proton, neutron-neutron, and neutron proton to form a nucleus. An imaginary particle named gluon mediates this force and there is a strange mechanical explanation for this interaction in the standard model with roles of elastic flux tubes.
Here,
We would like to explain strong force as an apparent attraction force between contents of the nucleus and also, between electrons and the nucleus.
There are two components of this force.

1. Mechanical Gravity push.
2. Magnetic force. (energy field interaction)

Let's describe them in a brief.

1. Mechanical Gravity push component.

As described in the section of the mechanical gravity push, the apparent attraction force between two masses is explained. It is described in that section that matter have porous hollow structure so, gravity push in that case is very weak. Nucleus of the atom has a mass relatively a condensed mass in comparison to element made of atoms or molecules. So here there will be a much obstruction to the flowing energy per square unite of the space. So, push pressure is much higher here due to dense mass in the nucleus at very short distance. This push works same as a push gravity.

For example,

If atom is considered as a ball of the size of the radius 100metre and there is a nucleus of 1cm at the centre of the mass with a mass of 1cumm or 0.01kg.

The effective gravity will be.

$$F = G \frac{m_1 m_2}{r^2}$$

$$F = G \frac{0.01 * 0.01}{100 * 100}$$

$$F = G \frac{0.0001}{10000}$$

$F = G * 0.00000001 = G * 10^{-8}$ units

Now if we assume that two mass in a nucleus in a direct contact as seen in nucleuses. Still, we would consider a distance between the centre of those two nucleuses equal to 1cm.

The effective gravity will be.

$$F = G \frac{m_1 m_2}{r^2}$$

$$F = G \frac{0.01 * 0.01}{0.01 * 0.01}$$

$$F = G \frac{0.0001}{0.0001}$$

$F = G * 0.00000001 = G * 1$ unit

Here, we have just roughly assumed the atom size and nucleus size for example.

According to modern science the real atom to nucleus volume ratio is 10^{15}. So, there will be huge difference in a gravity between between protons or neutrons in the nucleus than between two atoms.

So, this mechanical gravity push component of the strong force is also playing a major role in a formation of the strong force.

2. Magnetic force component. (energy field interaction)

As described in the section Torus formation, we can see that primary mass is in the form of torus. This Torus have a tangled-up energy field within and around it. Its field has a bidirectional flow of the energy as shown in the figure of the torus, and it has a polar flow also. When more than one such structures resides near to each other due to gravitational push, they have to adjust their position in a way that they least repel each other. There would be a tendency or a locking of this type of position due to directional energy flow in fields. There would be two types of interaction between energy fields, one is summing or uniting and another is repealing. The summing interaction would sustain because that would form nucleus, and repealing interaction would contribute to weak force as it would help to un-stabilize the nucleus. Now this uniting energy field interaction is due to polar form of the massive structure called torus. This interaction is between energy fields. So, this should be termed as magnetic force. When two torus units with such a magnetic force come in interaction, there would be a bond between them, that allows some movements. Oscillation of two tori would continue to occur. There would be a repulsive phenomenon also. And there would be limitation of the gravitational push also. Together that will create a weak force, and only when there would be a balance between strong and weak interaction, stable nuclear cluster would be created. There would be a field around that cluster also. Magnetic force component contributes the effective force between two such nuclear cluster. So, torus is the key component that creates a magnetic interaction in the subatomic particles. Two basic types of clusters could be there, tori and hollows. There could be a repulsion or attraction between two tori, but there could only be attraction between tori and hollows. There are both attraction and repulsion between two hollows. Its field would not allow them to stay nearer at very short distance. Due to lack of strong magnetic fields two hollow can't be locked strongly in a stable steady position with each other. A combination of hollows and tori would form a stable nucleus cluster at some specific combination only. Extra tori or extra hollows will create instability of the nuclear cluster. so magnetic component and push gravity together forms strong force. Explained factors that creates instability contributes to the weak force. Attraction due to field interaction at short distances is stronger than just gravity push. Thus, strong force is stronger than Gravity push only.

Weak force

Weak force and weak nuclear force:

This is a force responsible for the instability of the cluster.

Definition: when both push gravity and magnetic attraction force together can't counteract the kinetic energy, momentum of the matter cluster and repulsive field interactions which results in the radiation and scattering of the matter, the cumulative net force responsible for such instability and scattering of the matter is called weak force. When it is applicable to the nucleus of the atom it is called weak nuclear force.

So, weak force counter acts gravity, magnetic force at gross scale in case of large matter cluster while weak nuclear force counteracts strong force and magnetic force at a quantum scale.

As described in the chapter on gravity, when mass crosses certain limit, its random energy (internal random kinetic energy or field interaction responsible for repulsion) increases, and mechanical push gravity don't increase continuously as the mass increases. So, at certain point push cannot balance the internal repulsive forces and cluster gets un-stabilized. That disintegrates or get fragmented in smaller structures. This is a limitation of binding force, and that limitation contributes to the weak force.

When weak force works at sub-atomic level it is called weak nuclear force.

Binding force in nucleus has two components. 1. Mechanical push gravity component 2. Magnetic interaction.

Component one is pure attractive interaction so; it contributes to the constructive effect only. Here also limitation of this force appears at higher maximum mass of the nucleus. So weak force works here also in opposition to this mechanical gravitational push.

Component two is somewhat complex. The magnetic interaction can be attractive or repulsive according to situation and a position of the structure with magnetic energy field. In a specific locking position this magnetic structure gets united and locked in a position forming bonded couple. When thinking about adding more structure one by one, there are specific numbers that allow stable bonding, and others don't allow stability. When push gravity is enough to counteract repulsive magnetic force only then nucleus can be formed. Only tori can't form a nucleus cluster, there is a need for buffer units like hollows. Magnetic component is powerful in relation to gravitational component, so when this component creates instability, push gravity can't bound the nucleus in that combination. So unstable isotopes are possible at lower atomic number also. This shows an inability of both component of the strong nuclear force to create a stable nucleus, the repulsive force here is called weak nuclear force. Same as a strong nuclear force, weak nuclear force also has two components and an additional contributor which is a kinetic energy of the components of the nucleus. When unstable nucleus gets fragmented, it gets divided in possible stable subunits, this explains the radiation of certain particle and waves.

So, this is not a separate type of the force, but it is a limitation of the strong nuclear force. Science has not named similar thing in case of supernova which is responsible for explosion of the star. It is a similar force but not exactly same as weak nuclear force because weak nuclear force has a strong magnetic component.

Weak nuclear force allows decay and radiation even when nucleus is not huge enough to made it possible. This is because it is not always the total mass in the nucleus that results in the instability of the nucleus, but it is the imbalance between push gravity at nuclear scale and repulsive force between the subatomic particles or nucleons. When repulsive force between the subatomic particles becomes higher than a push gravity for that amount of the mass in the nucleus, that results in the instability of the nucleus. What creates repulsive force between subatomic particles! That is the spin of the subatomic particles or a rotational kinetic energy of the subatomic particles and magnetic interaction between them. random kinetic energy similar to heat also contributes. Protons are cluster of the subunits and have a rotational kinetics and a magnetic field and electric field that repulse another proton. So, two protons can't be held together by push gravity or mechanical strong nuclear force when that repulsive forces are not in a synchronization. But when neutron added to the nucleus, that adds mass to the nucleus, that increases the mechanical strong push force and that don't increase repulsive force in the nucleus. Neutron also provide a hide area for protons that further decreases repulsive force by preventing direct continuous interaction between two protons. So weak force doesn't break the stability of nucleus when neutron is there with two protons. Why neutron alone can't create a nucleus of the atom! This is because kinetic energies of subatomic particles are tremendous relative to their masses, so mechanical push gravity without a magnetic force alone can't form a strong bond in case of magnetically neutral neutrons. Neutrons have already achieved critical-maximum mass at that density level which won't allow a fusion between two neutrons. Neutrons may have nonmagnetic field that neither allow them to fuse with each other nor allow to be at locking position at particular distance. This means they have an attraction between them, but their field provides opposite repulsive force that don't allow them to make cluster. Neutrons can make cluster with protons, but themselves alone they can't make cluster because of lack of magnetic field around them. Magnetic field may be considered as a charge of the neutron, because they are directly related.

Electro-Magnetic force (curved field interaction):

Magnetic force:

(Topic 3 "Torus" in a chapter 5 can help to understand magnetic force better.)

Torus formation in Brahm is explained ahead in book in a chapter 5. Due to energy field vortex forms a torus with field. Such torus field is a curved field. This torus with field is a polar object due to direction of the spin of torus disc. Here in this book the word torus is used for a torus along with its field. So, such torus behaves as a mini magnet. Such tori with other non-polar nonmagnetic masses can unite in a cluster and able to form larger magnets when they are in a proper alignment. When two or more tori are in alignment and near to each other, their field gets merged to form larger single field. Torus is a spinning mass in a form of disc or intermediate shape between disc and globe with a field. That Spin counteracts gravity push and due to shadow of mass, the centre of such spinning mass forms a void of mass and void in a gravity push. This allows flow of the energy flux from above and downward also. This energy flux also gets attracted toward spinning mass by push of other energy flux. Means in a gravitational field shadow attracts and changes shape of flux passing nearby the mass. This will form a flow of deviated energy flux around spinning mass which forms a field of flux with power to push other objects and power to interact with other such fields. Thus, it forms a magnetic field and magnet.

So, these are primary magnets that creates interaction with other magnets due to energy field interaction. The force generated by such energy field is called magnetic force.

When such torus moves in the space, flux in the energy field gets replaced on the path of motion by moving energy flux of relevant places. So, there is a no interaction between Brahm and energy field of the torus because that energy field has its origin within Brahm, and it is not a permanent part of the Torus itself. Brahm forms an energy field of Torus, or a magnetic field of Torus. So, this field is a bending of the energy flux field in a Brahm due to shadow effect of mass, that forms a magnetic field. It can also be termed as a magnetic halo. Moving charge alone in Aether don't create an electricity but it has a magnetic field around it.

As described above, torus is a polar object. It has magnetic field. Two tori can have different interaction based on the opposing sides. If energy field is in synchronized position, that causes an attraction, otherwise repulsion if those fields are in opposing positions. Sometimes they unite side by side or in series to form tori cluster, which behaves as a larger unit of magnet and plays a role in a matter formation.

It is now clear that, bending of energy flow by spinning ring, disc or semi sphere can produce a magnetic field. The effect generated with this field is termed as a magnetic force. As it could be attractive in nature also, so that contributes to total attraction between masses, and it is a one of the components of compound gravity. This force contributes to strong force and weak force also.

Electricity

Spin energy and electric potential (charge):

When mass units arrange themselves in a cluster and form a nucleus, there are some possible shapes due to symmetry. Suppose we have magnetic iron balls. We want to arrange it in a smallest possible close pack cluster. Two will form a line. Three will form a triangle shape in one plane. Four will form three-dimensional shape with angular spikes. Some arrangement of 3d close packed spheres have spikes and some have relatively smoother surface. Few numbers would make a complete outer shell while few numbers would form incomplete shells.

One central ball and 12 surrounding balls would form a complete sphere with one centre and one shell forming two layers. This is made up of 13 balls. If number is 14 then the arrangement would be in 3 shells and outer most shell would have only one ball and that would form a spike. A sphere of 15 balls would have 2 in outer shell and a sphere of 16 balls would have 3 in outer incomplete shell. If total balls are 12 then there would be a gap in incomplete 2nd shell

This is just an arrangement of spheres in 3d. when nucleus cluster is formed, it involves arrangement of tori and hollows both. This type of nucleus forms a valency. A complete outer shell forms a stable and inert element like atoms of noble gases. A spiky outer shell forms a metal. Atom of such metal may have a spin momentum. When such atoms come in contact with other such metal atoms, it can transfer its spin energy to that atom nucleus. When atom spins, it enhances a magnetic field. Such field aligns other atoms by field interaction. Spin transfer and alignment is intertwined process. That result in a formation of series of magnetic structures. When spin is enough it leads to torus formation. This series of tori form a circuit that further sanctions more tori to be aligned and form a wider bundle of series. The transfer of spin momentum can occur in a series and thus electricity transmission occurs. Surface of Charged things have an increased magnetic field due to increased energy flow in tori and forming wider tori. Such increased magnetic field have attractive or repulsive interactions. This explains the rising or straightening of hairs when static electric potential is transferred to hairs via charged human body surface.

So, charge is an increased spin of tori forming the substance. Charge can create capacitance but not always create an electric discharge or electric flow. Electric flow is somewhat a different thing. Spin energy transfer just prepares the situation for electricity transfer, which is not always discharged.

Atoms can gain spin momentum by friction, or by motion in a magnetic field. That's why static electricity is there, and dynamo can generate a current. Charge transfer is transmission of such spin momentum. There is a no electron transfer, but a spin momentum of nucleus is getting transferred from some atoms to others which is called charge transfer. Not all type of material fits to transfer charge or create an electricity. Charged particle is a nucleus with high spin momentum which is able to spin other nucleuses with spikes. When heavy spin momentum is there than it can give friction damage to stable non spiky nucleuses, to stuck spiky nucleuses or to molecules also that results in heat generation and structural damage to material. If atoms are arranged in a mosaic in a way that allows them to spin and transmitting spin like toothed wheels, it becomes a conductive material. When atoms are packed with bonds or arranged as stuck cluster due to shapes, that don't allow them to spin and transfer spin like toothed wheels, that becomes nonconductive material. Sometime heat increases space between atoms and make them freer in case of semiconductors like silicon. Nonconductive material causes more heat generation due to stuck nucleuses in clusters unable to spin freely. That causes resistance to spin resulting in expansion of space between atoms or molecules and increase their random oscillation. This may result in EM radiation also. So, charges and charge transfer are such a simple thing but not the electricity is a simple thing. This also explains the concept of single charge, moving charge and charge-charge interaction as well as interaction of charge with magnetic field. This also explains why magnet slowdowns when thrown through metallic pipe. In that case, magnetic field or energy field of the magnet interacts with spikey nucleuses of metal and arrange them in a way that create interaction and motion of the moving magnetic field provides torque to spiky metal nucleuses and thus charges them. When they reach to an amount of maximum spin, it resists for accepting further torque so, magnetic field of the magnet get resistance and stuck there and thus slowdown. Furthermore, force as a relative motion may produce heat or structural damage to the materials. We can understand this phenomenon with mechanical sense, and we require no electro-magnetic laws to understand this. This is like many spherical toothed spheres or wheels interact by mingling with each other with spin transfer. The just difference is role of fields.

Please remember that particles in a nucleus have subunits and that subunits also have fields. So, when I wrote interaction, that doesn't mean direct interaction with close contacts of nuclei. This interaction involves field interactions. Moving energies have an ability to interact with other fields and masses. When one stable structure is there in Brahm, it has strong field and halo that don't allow direct fusion with other stable structures. It needs much energy to fuse two of such stable structures, normally they keep distance from each other. Those are not electron alone that provide a field as a shield, but it is a field of energy around nucleons.

So, these was the explanation of charge and charge transfer. You may imagine a simple charge transfer as an electric flow, but that is not the case. In conventional science, it is believed that electricity is the flow of the electron from high concentration or exited states to low concentration or calm state. But this is not so simple. If that was true, then electricity should be there from any open high voltage electric tension terminal to normal conductive material freely. There should be electricity in a wire joined to different terminals of two different batteries, but this is not the case. We know that electricity doesn't travels until circuit become completed. It is not like a high concentration of ions which get diffused to any low concentration fluid in a contact via permeable membrane. It is also not like a fluid at height in a gravitational field, that can flow to lower place by gravitational force. That electricity needs a closed circuit to flow, so it is more complex than conventional concepts.

Let's elaborate it.

We know quasars, they are massive structures, but smaller than galaxies in sizes. They are spinning entities and have an energy or radiation flow from its centre forming an axis. Neutron stars can also do similar things. The mechanism of formation of such energy flow is explained in the topic "Torus formation". At small scale, small tori can behave same as these quasars or neutron stars. When spin energy of torus reaches at high level it is termed as an exited torus. Such exited torus forms a strong magnetic field in its surrounding and axial energy flow appears during this excitement process. Such torus provides an influence that would align surrounding tori in a parallel and in series lines and transfer a spin energy. This line is not straight, but it is tortuous. This progress occurs in a series more effectively and it forms a structure like a neckless (mala) of pearls. When circuit is complete, means a path is a favourable, then this process gets repeated several times in that closed paths and sanction more and more tori to align in that stream, forming a linear flow of energy from the centres of such tori in a direction of the circuit. This flow of energy from the centre of the aligned tori along with field generated and spin transfer could be termed as an electricity. We can now understand that why there is a need for a closed circuit for electric flow. This can also explain the phenomena related to faraday cage. In a faraday cage a closed circuit forms a shell-like path in which tori are aligned, so they are not able to transfer electricity to other direction toward a sequestered area within that closed shell.

These was the explanation of the electricity and charges. These explanations are in support of the proposed Brahm model containing tori and push gravity.

This model can also explain light emitting diodes and electricity generation by photo electric cells.

Cathode rays can also be explained simply with this Insight.

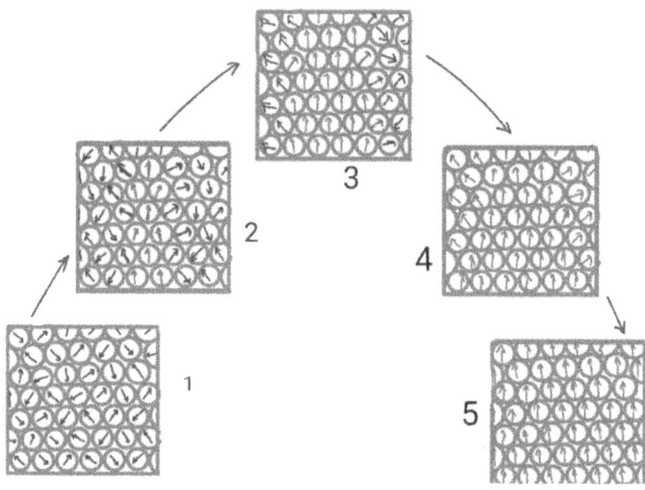

Figure 4.4 showing serial alignment of tori forming circuit.

Spin energy also explains acid battery function and storage of electricity as a spin of protons or ions. This also explains corrosive nature of acids. Acid battery shows relation between spin of ions and electricity.

5. Structures in Brahm

As explained during the explanation of the gravity following structure are formed by the formation process powered by the same environment that creates push gravity.

1. Nakkar (solid)
2. Hollow (polo)
3. Torus
4. Mix cluster
5. Brahmand

primary mass in Aether is a result of cluster formation of resisting things in Aether or waves in Aether by push gravity. gradually they unite according to their field interactions and form the elemental matter. When there is less than enough matter then it just forms a cluster with no void at centre.

When gathered mass becomes enough then mass itself interrupts the pushing flux from penetrating to centre. So, here mass in the centre lacks push effect from surrounding and so, lacks gravity. Mass tries to get disintegrate in subunits or energy at that void area in centre in absence of the effective push gravity. it is explained in a topic Galactic cycle.

When mass is moderate then the central void in the effective push gravity would remain incomplete, and mass disintegration would remain partial at that centre. When mass is enough enormous then void would be complete.
Two type of toroid structures of mass are possible.
Disc like structures and oblate spheroid structures.
That depends on the spinning kinetic energy and on the nature of the substance forming mass. More spinning rate and polar units in substance leads to disc formation, and less spinning rate and non-polar substance leads to spheroidal structure formation. Spinning a spheroidal structure to very high rate, may

transform it in disclike structure. Disc structure has a void in centre and that forms a torus, while a spheroidal structure doesn't form a strong torus and lacks heavy encircling energy field around it.

Examples of spheroidal structures at large scale are planets and stars. Visible Structure of matter smaller than planet may not achieve a globular spherical shape due to less mass and less gravity working on it. But they fall in this category only.

Up to certain limits of mass, globular structures increase in size while adding mass to it. When mass is in access then the structure becomes unstable and either it gets exploded or starts to lose mass. There is also a limit at what it becomes self-luminant. At some critical mass gravity becomes intense enough to create a state that will form light. So, up to some mass limit it forms planets, Brown dwarfs, red dwarfs and stars as the mass increases. Brown-dwarfs are considered as failed stars and Red-dwarfs as small stars.

When such masses are formed, they feel attraction toward one another. Plenty of such structures may form a large vortex, whirlpool like a structure. Which may gather mass at periphery and disintegrates mass in the centre. Galaxy is such type of structure.

Smaller disclike structures are quasars which has visible jets of energy because it has incomplete voids at the centre. While in full formed galaxies the jets may not be visible normally, because of complete or near complete voids in the centre.

1. Nakkar

If we accept aethereal-cycles, then as described in the chapter "Honi", not all solid continue to get blasted or disintegrate in smaller fragments. Some of them gain a support of the push gravity generated during process of disintegration, which stabilize them and thus they escaped from the process of disintegration. This is a thought experiment in a situation without any push gravity which is unrealistic, and it was just imagined understanding ultimate consequences of imagined mass in nothing. That's why we were unable to describe that mass further at fundamental level, The real cycle may be different type with continuous process rather than such isolated prime big bangs. So, if we try to explain that blast in a real situation, we have to accept that there could not be the absence of the push gravity, but it may vanish to some little amount that can't hold the masses with its pressure. Solids imagined in that thought experiment becomes a part of cyclic process. If we want to imagine the root of these solids, they could be the remnant of previously formed mass or Aether during aethereal cycle. We won't go deeper now to know about its composition, for us it is sufficient to know that it is a stabilized mass which has a property of mass, which can be acted upon by push gravity, which can carry momentum, which can cause impact with other masses. So, these solids are remnant of the past old structures. The word for that in Indian language is "Bhut". Aether if disintegrate due to lack of supporting push gravity than it can also behave like a disintegrating mass because it is made up of mass indeed. Cluster formed with such bhuts is called Nakkar. Nakkar is a word in Gujarati language for solid. I used it to differentiate it from other type of solids. Elemental matter can also have a state of solid. So, for unique identity the word Nakkar is used here to differentiate it from other types of solids. But these are intracycle existences which fits in a cyclic Aether state only. For a steady state, continuous formation of mass differs from this as explained in the topic "A new mass formation".

When we talk about the vortex of waves in Aether, that forms a torus structure, we must have a question in a mind about a material pushed together. What could be that material, which is pushed together by push gravity which results in vortex formation? That's why Nakkar answers this query partially here. It is the answer for fundamental solid unit. Its size is decided by effective push gravity at particular scale. Larger solid will be disintegrated in smaller particles and smaller will be gathered to form solid of particular optimum size. So, at one scale, two type of mass would be possible. One is a primary Nakkar of optimum size and another is a cluster of smaller Bhuts as a stabilized globe or Torus.

Primary Nakkar is termed as Bhut also as it is thought to be a remnant of the previously formed non disintegrated portion of the previous Aether or mass before multiple bangs. Bhut means what remains after vanishing of previous existence.

The word "Swayam-Bhav" literally represents two meanings for the existence. 1. that is a self-stabilized and self-existed with self-support without need or help of anything else and 2. which is formed itself. Though it is not imaginable, but something which is not nothing actually represents such thing. Torus, hollow, Aether, mass etc. are also represents such thing because for the occurrence of these things there is no need of active manipulation or will. They are product of the natural process. They are self-stable and there is a no need of support for them when they are in a Brahm.

A new mass formation.

Brahm is imagined as Aether with waves within it. When plenty of waves per volume of Aether are moving in all directions, there is a chance of collision between waves. They may collide at different angles. Waves can pass over each other while motion in the Aether. Most of them may not collide but the probability says that there must be some chances of collisions that may result in some consequences. The angle between colliding waves may result in different outcomes. Stright collision may result in a disruption or overriding. collision at a smaller angle may result in direction changes. At some perfect angle they may form a binary system spinning around a common centre. If we think of waves within Aether, then they have ability to stay closer to each other and form such couples and clusters of waves. Such binary system is larger than average moving single wave. Their kinetic energy is transformed in their Spin which resulted in slowing down of their linear motion. It will become a spinning stagnant resistance in a Brahm which would be acted upon by flow of waves of tides soon. Further contribution to this binary structure will form larger structure if all goes well. Process gets repeated until it forms a good cluster. Now such big clusters are newly formed primary masses. We haven't considered primary moving aethereal wave or tide as a mass here, because they are the units which forms a push gravity and don't feel push gravity until they get fused together. So, mass here is larger than this primary unit of tides. Further consequences of such newly formed masses are similar as described ahead. They form various aethereal structures and forms particles, atoms, molecules, matter, clouds, nebulas, stars, platens, galaxies etc. Chances of formation of such masses are there at every point in a space, but there are increased chances within a field of torus due to deviated or bent path of the energy flux. Such newly formed mass enters a galactic cycle and passes through its journey and becomes a part of the non-living and living things and it may move toward the centre of the galaxy where it gets annihilated in separate individual waves or in smaller clusters.

2. Hollow

This is a 3d structure, as described in the gravity section and above description, mechanical push gravity can lead to cluster formation. There is a gradual decrease in the reach of the penetration of gravitational pushing flux as the mass increases and gathers further mass progressively. This will create a cluster described here as a hollow. It has a crust where there is maximum pressure due to mechanical gravity, but beyond that crust, there will be a gradually less dense area with high temperature and high local kinetic motions. Outward from the crust there would be an area of equilibrium, where there would be continuous exchange of mass and energy, forming a corona if this is a full-sized hollow structure. Maximum pressure area is not prevalent on surface but exists deeper from the surface due to combined force of gravity forming flux and direct pressure by outer pushing mass. So up to certain depth pressure increases, but after that depth gravity forming flux gets interrupted, and that would result in a less gravity area where pressure due to direct gravity becomes low and mass tends to have more kinetic energy and heat.

This structure can stay in this position when there is no optimum effective spin of the material forming it. If the material forms this structure while having an optimal spin, then it may result in a structure Torus. This Torus can be formed only when there is an optimal spin. If the spin is intermediate between required for the Torus and Hollow, then it may result in a formation of the intermediate structure that can acquire an intermediate shape like an oblate spheroid.

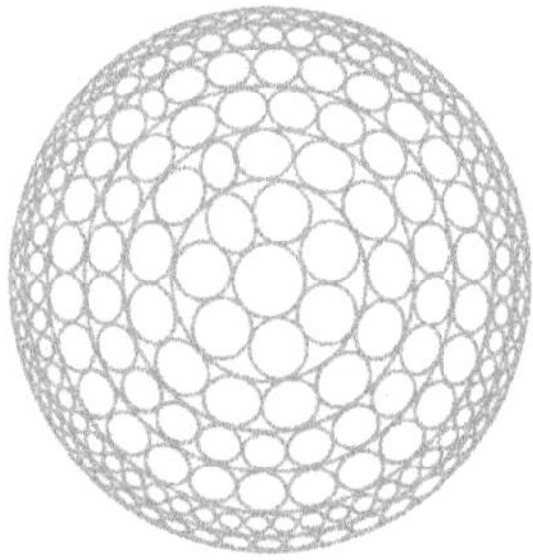

Figure 5.1: Showing the core of a hollow.

If we think of masses at large scale which are categorized as hollows, then those are stars.

3. Torus

As described above in description of the hollow, when material forming hollow is having an optimum spin then it can produce a centrifugal force enough, that can flatten the distribution of the material in near one plane and result in the formation of the structure Torus. A magnetic field generated while spinning also contributes to arrangement of the material in one plane via mechanism of the field interactions. There would be a different size of such Toruses (Tori). If mass in the torus would be enough to obliterate push gravity forming flux sufficiently, then it would produce a void at centre, and this would result in complete torus formation. This void centre would be an area of minimum gravity, where there would be a gravitational void formation, and it would allow energy flow from two sides, above and downward. This bilateral energy flow may have interaction in this narrow void.

Figure 5.2 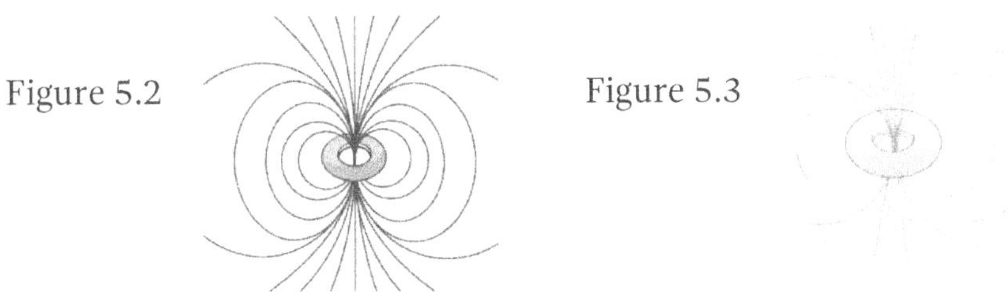 Figure 5.3

This would result in a flux-field formation that would have a path as shown in the figure above and it would contribute to magnetic field formation and thus completing formation of the Torus. This is the single and a complete Torus with positive and negative poles. It has a polarity depending on the direction of the spin of the material disc. You can compare this Torus with structures like galaxies, quasars when it is a complete and near complete form. Solar system is nearer to the type of hollow. Small Planets are matter clusters with a mass of less-than-optimal mass to create a hollow structure.

Toruses have a strong energy magnetic field with the ability to create an attraction and repulsion depending on the polarity of Toruses. Toruses are basic magnets and provide a base for magnetic interactions. Hollows are magnetically almost neutral structures and have more affinity for gravitational attraction only and no magnetic directional motion in magnetic field. Still, it repulses other hollows due to radiation fields. Hollows are comparatively dense structures and tori are comparatively less dense structures with plenty of non-materialistic voids.

Torus Formation:

A material in Aether is any structure in Aether that resists energy flux. Such materials are pushed together by energy flux and if not obstructed then form a cluster. During this cluster formation, that material get condensed, in this process when material spread over large area get condensed in a smaller area, due to conservation of momentum or angular momentum it achieves a spin. This angular momentum due to spinning causes some centrifugal drag, that counter acts push gravity. When mass becomes enough, that can interrupt enough push gravity forming flux from penetrating toward centre and create a void at centre. Materials moving in orbit are made up of polar masses. Due to the combined effect of field interactions and centrifugal forces this material forms a vortex or a disc like spinning structure. The material moving in orbits forms a disc like structure with fields, a Torus, which creates a bending of energy flow from the centre of that Torus disc. This bending of the energy flux suggests that energy flux itself has a nature to be pushed by energy flux and can have a deviation from a linear path around the mass due to push gravity. This also points to the material nature of the energy flux. So, basically the nature of gravity, which is defined as an external push force, is responsible in such Torus formation with its fields. If we accept gravity as a fundamental attraction between masses, then it won't create a void at centre, it won't explain a central axial energy flow or annihilation of matter at the centre of galaxy. So, push gravity is the ultimate key behind this model. The disc has a spin and that affects the direction or path of the field, that forms twisted patterns in field, that plays a role in interaction between two tori. That also determines polarity of poles.

4. Mix clusters

Toruses and hollows are aethereal structures, and they were explained ahead. If we inquire inside the atom, we will find a nucleus in the centre. That nucleus has an atomic number, that defines it as an atom of particular element. Protons and Neutrons are well known subatomic particles. Protons have a deviation of path of motion in the magnetic field, while neutrons lack that quality. From the nature of aethereal structure, we can assume that proton could be toroid structure or cluster formed with tori, while neutron could be the hollow like structure or cluster of tori with no net directional magnetic field. Torus has a strong magnetic field while hollow has a radiation field. Because of their such properties, some unique combinations of these are possible as explained in the strong force chapter. These combinations form various types of nuclei.

This nucleus can result in smooth form when the number of nuclear particles is such that it can complete a shell of hexagonal symmetry. If the shell is not complete, then it will result in incomplete shell forming a spiky surface. Depending on the rate of spin of nucleons, it would form a field. The spin of the whole nucleus can form a toroidal like structure. This type of spinning nucleus can be termed as charged nucleus, that has a spinning potential energy ready to be transferred to similar less spinning structures. Such toroid nuclei are charged with ions. This is an example of static electricity and charge. Spin of nucleons or spin of whole nucleus or both may contribute to such potential energy due to spin. This may explain why some elements are conductors of electricity while others are not, because of favorable structure of their atoms that allows spin transfer due to combinations of spikes.

You can see here that this is just a logical derivation based on conceptual thinking only. There is room for improvement if we try to fit experimental data in this model with appropriate changes and adjustments. But this model provides enough structures with their fundamental origin and characters that can create a subatomic system.

It is very difficult to imagine a real situation at this level, because there may be so many possible combinations of aethereal structures. Their stability varies depending on relative strength of push gravity, mass, magnetic fields, spin energy, heat energy, etc. So, at the level of atom formation or subatomic level, retrograde approach to understand that structure as a combination of basic aethereal structure may be help full to understand that structures. At present sense to understand mechanical interaction and mechanical behaviour of aethereal structures in a Brahm is limited but if there will be a critical thinking and experiments on this, it will yield more knowledge and understanding about subatomic environment.

Elemental matter

When we are talking about elemental matter, we must have to describe an atom. Atom is a basic unit that forms an element. We know that there are many types of the elements categorized in periodic table useful in chemistry as well as in physics. Depending on the category or groups it shows specific characteristics also. There is an importance of so-called free electrons in the outer shell of the atom of that element also. All elements have different atomic mass also. Number of protons and neutrons are assumed in the nucleus of the atom also. Electrons are assumed orbiting that nucleus. Positive charge of proton, neutral charge of neutron and negative charge of the electron are also determined from the findings of various experiments, particularly deviation of the path of moving particle in a magnetic field is helpful in determining charge of the particular particle.

What explanation and model could be provided to explain these things, in a scope of Brahm theory? Actually, a better model than a standard particle physics model could be provided with this theory. Charge, spin, sub-structure, explanation of behaviour, radiation, decay, etc. could be explained with more fundamental insights. But that all could be provided as logical assumptions only. Correlation with experimental data can provide a proper insight.

Let's begin with logic only.

Let start with charge and electrical activity.

Charge is a characteristic of the particle determined by its behaviour in a magnetic field. If it gets attracted toward a negative pole of magnetic field, its charge would be considered as a positive, if it is attracted toward positive pole of magnetic field, its charge would be considered as a negative and if it won't show any deviation in a magnetic field, it's charge would be determined as a neutral charge. These are the conventional concept about the charge of the particle or ion.

To understand these things in relation to Brahm model, first we have to understand the magnet and the magnetic field.

Magnets:

I have explained the primary magnet as a torus in this same chapter above. Field of the energy flux is also explained. Such torus or cluster of unidirectionally parallel aligned tori forms a magnet. But magnets are not same thing as static electricity, because in static electricity, spinning units have potential to transfer their spin to receiver units. But in case of magnet, it is not so, magnets are not always electricity conductor. Those are aligned tori embedded on a base material that fixes their orientation and thus forms a united field and a large magnet. Thus, a combination of ferromagnetic metal and non-ferromagnetic metal used to form permanent magnets. Single ferromagnetic metal can also form permanent magnet due to its property as a solid. Magnets have a surrounding field of bent energy flux. This field is a bidirectional field and not a uni directional field, still magnet shows poles. Positive pole and negative pole. The direction of the spin of the disc of tori determines poles. When opposing poles of two magnets near to each other have synchronic spins of disc of tori, it creates synchronic twists in energy flows. When those spins are opposite, it would create a non-synchronic twist in energy flows that causes repulsion. Charges are also small magnets. During interactions, some gives of spin and some gains spin, and they exert behaviour of deviation in magnetic field according to their spin rate. Imagine these things like a fish trying to climb a water flow from above to downward obliquely. If speed of fish is more than speed of water flow it will move upward and climbs flow, if its speed is less than the speed of waterflow then it will move downward with the flow. Neutral things lacking net twisted torus fields and thus lacks interaction with magnetic fields.

As explained in the topic clusters, various nucleuses are formed. Their size and spikes define their nature as an element. Nucleus is the atom and atom is the nucleus. We are not talking about electrons orbiting nucleus. Spinning nucleus and nucleons within the nucleus forms a field that contains electrons. Magnetic fields of the nucleus and nucleons form a field for the atom. Which enhance the physical boundaries of the atoms beyond physical boundaries of the nucleus itself. There is nothing like electron shells of electron orbits. Nucleon's field is the area of energy field around nucleus. That is the area of the nucleus what forms its shape and character. Nothing is there to balance the positive charge of the protons. It is a spin rate what determines a charge of an atom as positive or negative. When proton alone has an increased spin, it forms magnetic fields and behaves as a charged particle. When it has less spin, it lacks such torus like strong magnetic field and behaves as a hydrogen atom. Such charged nucleus like a hydrogen ion can have power to cause a physical damage to other elemental matter. It is revealed in case of corrosive nature of acids having high concentration of such excited nucleuses. An element like aluminium has poor structural integrity compared to other strong metals and have receiver spikes on its nucleus surface renders it vulnerable to damage by acids. As described ahead in the topic electricity, Battery works on this principle also, where electric charge is stored in a solution as a spin energy of the ions or elemental nuclei.

Fields of the nucleons don't allow the fusion between two atoms. Push gravity magnetic field and structural flexibility works together to form chemical bonds, that determines state of the element or compound. Affinity of different types of atoms toward one another due to shape and structure of nucleus provides specific state to the elements. This also provides chemical and electricity conducting qualities.

5. Brahmand

"Brahmand" word is used synonymously for the word "Universe". There are few imaginations about the shape of the universe and boundaries of the universe. Observable universe is just a part of the whole universe that is visible to us. Brahmand is the word that has a meaning of the globular or oval shaped universe with distinct boundary. Brahm mean Aether and "And" means an egg. So, Brahmand means an Egg of Brahm.

In this theory I have described a distribution of the matter in pockets and its complex arrangement due to torus formation. In this distribution there is a level where stable hollow is arranged in a lattice which forms a substance aether with resilience and elasticity. The unit of this substance aether is a hollow with stable structure that has a field around it, a crust like shell and then there is a cavity with less dense material. That material is formed by clusters of tori arranged randomly. These tori are galaxies with large toroid fields around it which arrange them in clusters.

This brahmand must be either moving or vibrating. How it prevents internal structure from those jerks and vibrations and provides a piece that allows a formation of tori within its cavity? Inertia should work against this. Answer is in a motion scale. For a motion at the scale of Brahmand there are very fast motions at the galaxy levels which provides a strong opposing force and allows them to accommodate with Brahmand. Brahmand also provides a Nobel thermodynamic system that makes this system a closed system.

6. Mass, Energy waves, Flux particles and Inertia.

Mass and Matter

What is the matter? The literal meaning of the word as I interpreted is "That's why it remains sustained". Means it is a sustained stable thing with explainable reason behind it.
So, it is the answer to all fundamental questions related to existence starting with what and why.
The origin of the word must belong to one of the most ancient languages or most ancient civilizations. The depth of the ancient theories or puraan can be imagined with this word having a deep-seated meaning. Well, we are not going to accept it without satisfactory explanation of nature of matter that must match this meaning.

Now there would be a description of the matter just by logical concepts that must obey the laws of Newtonian motion mechanics and thermodynamics. One can understand the gravity as described above in this chapter. Let me fit that concept in the observational data. If gravity is really working like mechanical push, and if we put multiple large masses in a line, then what should happen? Does the mass at middle of that series would feel the gravity? yes it would feel the gravity, because in a series of mass the gravity push forming flux will not get obliterated from other remaining sides. Other factor working here is that mass is a porous thing and not an absolute solid. So, obliteration of energy flux will be very little. The push is generated by just a small fraction of the total energy moving in a Brahm. So even a very long series of the masses will not completely obliterate the energy flow generating a push effect.
Can we imagine how much porous the atom of matter is?
It can be roughly calculated based on observations. We know that atom has a space within it. From the experiment done by Rutherford we know that a nucleus is a very small compared to whole atom. If we put two atoms side by side as near as possible, even then the nucleus's proportion to the total space acquired by the atoms is just less than a millionth of part. So, the matter we

know as element has a too much space within it. If we see at large scale, such matter makes stars, stars make galaxies. In a galaxy there is too much space between two stars. There is a factor other than angular momentum that don't allow all stars in a galaxy to collapse into one mass. Stars remain scattered in galaxies. Most galaxies are spiral galaxies, and the observed mass made it porous object and not a solid cluster. Around the sun, there is a Helio sphere. Galaxies also have such fields, which is formed by energy flowing and encircling galaxies in a manner shown in a structure of "Torus". Galaxies are also scattered in the cosmos, and they have size limits also. If we see the cosmos as a large-scaled matter, then we can understand that it has a very much free space or voids within it. So, these voids allowing the gravity to work how it is working. In other words, these voids are there because push gravity would distribute matter in pockets only with space between them which further forms a large pocket. These pockets are not simple globes due to magnetic property of the tori that provides a specific pattern due to field interactions. Those are not just fields that have placed stars apart from one another and same in the case of galaxies also. A long process that put them in that calm state is full of events powered by forces generated with push gravity. This calm state is achieved after many interactions over long time span. Fields have played a role, but they are themselves not capable to prevent collisions if any object have a motion toward other such large object in case of structures like Galaxies or stars.

Now we will think about the area having minimum gravitational push due to obliteration. Where such area can be possible?

Yes, it is possible at the centre of the very big mass cluster, like a very big star. But we can't inquire or observe it due to inaccessibility to that deep area. but we can observe similar area at the centre of the galaxies.

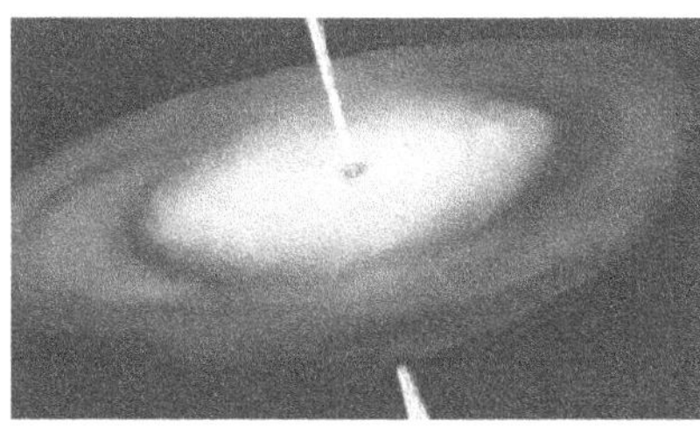

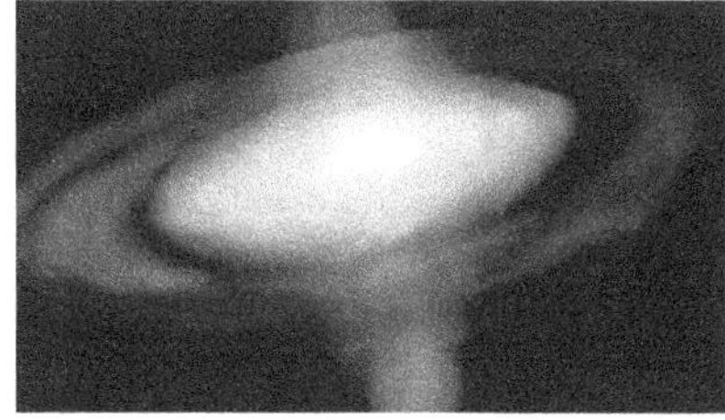

Figure 6.1   Figure 6.2

As shown in figures 6.1 and 6.2, due to tremendous mass in a galaxy and around it's centre, the area at centre is void for the gravitational push. In the absence of this pressure, matter scatters toward that centre and get dis-integrated toward centre. The primary vortices in Aether behaving as a primary matter stays stable due to gravitational push. When the gravitational push vanishes in a centre, then all primary vortices in Aether and their clusters get dis-integrated and clustered units becomes free. In that area energy flow don't have access from the sides due to obliterations by the mass in a galaxy, but it has access from above and downward sides. Energy flaw moving with maximum speed from that area interact with that disintegrating matter and creates flashes of lights or other radiation when galaxy is not large enough. From these observations we can observe the mass needed to create a pressure free field that allows matter to get disintegrates or annihilated completely. Quasars are the objects observed with such flashes of lights. But due to its small size compared to galaxies, that are unable to create enough gravity free centre, and that results in partial raw annihilate matter leaving fountain formed by interaction of energy flux with that plasma.

So, these observations support the imagination of matter as a cluster of vortices made of waves in Aether stabilized by push gravity. Here we can also get the assurance that stable form of matter can't get disintegrated until and unless it is placed in such voids having no or negligible gravitational push. So, matter don't have a half-life if it is a stable isotope. Basic fundamentals of the matter are also described ahead in a chapter 5.

So, units of matter are complexly arranged vortices of waves in Aether medium with torus discs and field around it due to energy flow in Brahm. This primary mass stabilized primarily by mechanical push gravity generated due to obliteration of energy flow in Brahm. It is described in the chapter "Torus" that vortices are formed due to push force that forms cluster of smaller vortices. So, matter is a compound made up of multiple tori or vortices and hollows which are made up of pairs and cluster of unit corpuscles or waves in Aether.

What forms the primary mass? We have seen that vortex is formed in Aether, because push gravity brings together some masses in Aether. If we imagine that mass as smaller vortices, than again question arises in the mind that what forms primary mass, that may not be necessarily a newly formed vortex of waves in Aether? Here comes the one answer from the chapter "Honi". During the process of disintegration or blasting of mass as shown in chapter Honi, Bigger mass disintegrates in smaller and smaller fragments. Not all fragments continue to disintegrate in smaller ones, but few remain intact and stabilized due to formed push gravity during that process. Those remnants play a role of primary solid, which forms vortex, Nakkar and hollow like aethereal structures and contribute to the formation of aethereal vortex and nuclear cluster.

As described in the topic "New mass formation" primary mass or a resistance in Aether is formed by unification of waves in Aether.
So, mass is either a primary cluster of aethereal waves or a cluster of smaller masses brought together and stabilized by compound attraction made up of pushing flux.

Mass as a cluster of aethereal waves

If we consider time dilation as a true phenomenon then we could get read of the primary vortex as a couple waves or clustered waves within Aether Substance. Why this is so and how primary mass as waves in Aether explains time dilation? Waves in a medium travel with a constant speed determined by elastic property of the substance. If primary mass is a clustered waves within aether substance, then its component waves must have a finite constant speed in relation to aether-units. Only circling or orbiting waves can form mass by forming waves cluster. Now primary waves in such wave cluster have to orbit the centre of cluster continuously. When there would be a linear motion of such mass cluster, waves have to travel in two directions. One in the orbit and another in the linear direction. So, as fast waves move in a linear direction their orbiting speed decrease. To achieve this, waves acquire deviated path which is longer to complete orbit in comparison to original orbit path when cluster is not moving in a linear direction in relation to Aether. Time depends on the length of the orbital path on which waves have to complete the orbit. As time for completing orbit increases, time slowdowns for the wave cluster or mass. Thus, we can understand the mass as waves within Aether. Aether is also made up of mass and push gravity forming flux. To find a root of mass, we can go to Aether and its waves. To describe the root of Aether, we can explain it with arrangement of mass within a push gravity environment by lattice and elastic substance formation.

Antimatter:

Depending on the orientation of the spin of the tori counter mass or a counter particle can be possible. Previously abundance of matter and particles don't allow accumulation of anti-matter or anti-particles. There is a no difference in the stability of antiparticles in a matter free vacuum. Particle anti particle interaction can result into aethereal waves like formation of Gamma rays or photons generation. It is like unwinding each-other due to interaction resulted from an attraction toward each-other. No exitance of anti-photon Favors a possibility of photon being an aethereal wave and not a maas or particle. Galaxy is a progressively increasing structure and it has a toroid field. This field don't allow a formation of antimatter.

Energy waves

Energy waves are nothing but a wave as a serial deflection of the unit cells in Aether matrix which is behaving like an elastic medium and a magneto gravitational substance. So, this are aethereal-waves. The closest possible arrangement of globular objects is a hexagonal symmetry. So, waves are moving series of deflections in this system of symmetry. The word symmetry has a meaning of looking for friendship within. The units in the system are already saturated with energies and also aligned tightly with mechanical push gravity and locked in positions due to magnetic interaction of vortices. Energy flows are coming from all sides and going to all sides. So, units are continuously vibrating and acquiring space enough for these vibrations. Cells when get deflections, they pass it according to stroke, which it has received from the stimuli. Harder the stroke, wider the wave and it may decide its frequency. Wavelength is in inverse proportion to the wideness of the wave. Because the surrounding cells are saturated with energy they won't absorbs energy of the wave, so the wave won't get scattered. But it will move in one direction. There is possibility that there may be a pair production during the formation of the wave, and both have complementary characteristics with each other and opposite direction of propagations.

Tapering of the waves or Wave transformation

As the energy waves travel large distance, the interaction between them and the medium carrying them is there. There is a possibility that the waves may lose some energy, but I want to describe something else here, and that is not due to absorption of the energy alone, but that is due to transformation due to tapering by resilient and elastic surrounding aether units.

First, we will observe the following experiment.

In this experiment a heavy small wheel of the metal is used. If we put it on the solid flat smooth surface on the earth, in presence of the gravity and spin it in oblique position with applying a torque with fingers, we will see that it will be rotate on the surface ground. The max height of its edge in the beginning of rotation will be highest, gradually it will decrease, but that will not be only due to loss of energy due to friction or heat loss.

With decrease in the max height of the edge of the wheel, the speed of wavey rotation will increase. It will become so fast near the end that it will become difficult to count manually without help of special equipment. What increases that speed of wavey rotation? That is due to gravity. Its height in an initial phase has a potential energy due to its higher position, that potential energy gets converted to wavy rotation as the event proceeds. That gravitational potential energy is converted in to increased speedy wavey rotations, and wheel moves very faster in the end. But height of that wavy motion decreases gradually because gravity pushes two mass closer, meanwhile rotation has to accommodate the converted energy in a wavey circular motion. So, speed of circular rotational motion increases in rate.

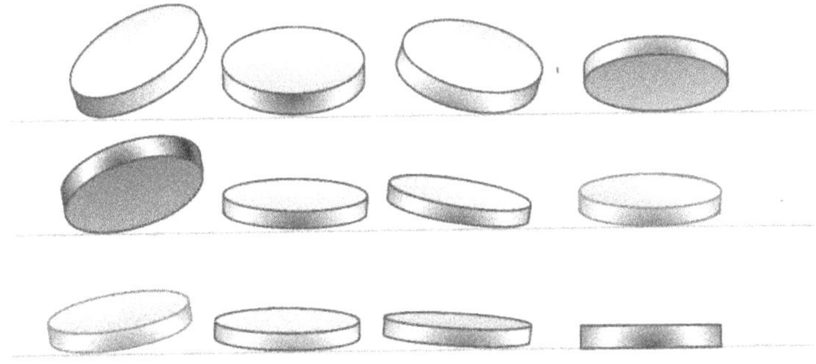

Figure 6.3

This is called motion transformation or motion tapering. Similar things occur with the light wave or any other aethereal wave in the space where a tapering or compressing force due to elasticity of Aether or effect like a push gravity aethereal wave gets tapered over time and it becomes narrow gradually. That looks like a decrease in the frequency and increase in the wavelength, this transformation mimics red shifts to some extent, and is mistaken for Doppler effect or astronomical red shift what gives one of the two bases for expanding universe theory.

Figure showing such an effect.

Figure 6.4

This is just to explain the phenomena, actually it takes many more cycles to gain such effect, and in case of light wave width decreases and length of wave increases. The change is not linear with decrease in the energy content of the wave, but before vanishing of the energy, this transformation occurs due to nature of Aether medium propagating that wave. It is not determined here whether aethereal waves lose energy while propagating or not.

This is the explanation for apparent astronomical redshift. This was the explanation of energy flow as aethereal energy wave. But the possibility of light as a particle has not been ruled out. How can a Particle have a property of the wave? Is that possible? It is possible. We know that in double slit experiment electros show wave nature also and wave function is applicable to every quantum object. There is another one possibility of mass moving like a wave.

Fish Projectile

Imagine Brahm as a gravitational environment similar to model imagined by Nicholas Fatio. Means there is tremendous flow of particles from all sides to all sides in empty space. Their concentration is so much that it doesn't allow the matter to feel empty space at all and stabilizes matter by strikes from all sides forming a push gravity and strong force.

Now imagine a cluster of tori with shape of the fish. This is stabilized by push gravity and magnetic force. This fish has a tail like projection. By somehow the force is applied to the tail, and it starts vibrating. What will happen next? Yes, it will move due to propulsion done by that vibrating tail. But there is a push from all sides, that will potentiate that propulsion tremendously and that fish will become fish projectile and will move with very fast speed. It will form a linear moving thing with wave pattern formed by tail. It will have its head ahead of wave.

The frequency of vibrating tail will become frequency of the wave formed. As shown in the example of the heavy metallic wheel in above description, the vibration of the tail will be transformed by push gravity, and it will gradually become faster and narrower showing tapering of the waves produced by the tail. This will also mimic redshift.

So, wave transformation phenomena equally applicable to particle form of the light also. So, possibility has not been ruled out that linear waves could be a fish projectile only and normal waves can be dispersing waves only.

Imagining a photon as a torus-particle also shows such tapering effect. While moving in a Brahm and gravitational environment axis of torus-particle would oscillate to achieve stability, the width of such oscillation would decrease gradually, and speed of its occasion would increase. This also suffice the nature of light and tapering phenomenon.

Inertia

This topic is mentioned in introduction, let's see in detail.

In physics inertia is referred as the characteristic of the matter which shows a tendency to do nothing or to remain unchanged.

Newton's first law of motion represents the inertia.

Inertia is a property of matter by which it continues to stay in its existing state of rest or uniform motion in a straight line, unless that state is changed by an external force. The Latin word inert is in the centre of the word inertia. Inert means unchangeability.

Newton's first Law of motion:
"A body continues to be in its state of rest or of uniform motion in a straight line unless an external force is applied on it."
This is called a law of inertia. This nature of the body is called inertia.

This means, when you add a push to an object by applying a force, then it will transfer an energy as a momentum to that object, which will be retained by that object until further interaction with another object or force. This means the object by itself don't lose a momentum in a vacuum or space.

In absence of a gravitational field, moving object continue to move in straight line. In presence of the gravitational field, object get attracted toward the mass creating that gravitational field and final tract will be resulted according to vector addition operation between momentum and gravitational force.

Momentum

It is an expression of the inertia as a product of object's mass and velocity. When mass gains velocity its mass increases due to incorporated energy(tides) with the moving mass in addition to velocity.

Relativeness

In a modern physics as the law of inertia states, there is no variation in a behaviour of an object at any speed after addition or subtraction of the momentum. Relative mass is applicable for relative motions only. This means relative motions are relative and there is a no universal frame of reference that can determine an absolute motion of any object. One cannot judge about any object, if it is a steady or moving object until its position or state is compared to another object or with many objects simultaneously. When there is an application of unidirectional force to any object, it will get an acceleration, that will increase the velocity of that object and will increase its relative momentum. There will be increase in relative mass also according to special relativity. Thus, with this concept we cannot determine an absolute momentum or an absolute speed of any object. This description is done within the limits of Eistein's relativity.

The above description is within the context of Einsteins relativity known as a special relativity and general relativity. Those don't fit in Aether-Brahm theory without modification to accommodate non-moving reference frame Aether and dynamic reference frame Flux within a Brahm.

Queries about inertia

Inertia is a memory of momentum in an object. Let have a one thought experiment.

We move a 1kg object and it crosses first and second points in a space with uniform rectilinear velocity. Suppose it acquires a velocity of 1000 meters/sec. A distance between first and second point is a ten thousand meters. This simply means that it will take a 10 sec for the object to travel a distance between first and second point in a space if we neglect time dilation. Here we assume in a classical physics that space or vacuum without gravitational or magnetic field is like an absolutely nothing and provides no interaction or resistance. During the process while object crosses the first point and travel towards the second point and reaches the second point, there is a no work done during that process. There is a no interaction with that object. It was the only isolated object cut from the surrounding world in that area of vacuum. How could that object maintain a steady velocity when it was moving in that area? How could it maintain a straight path? Why it took time to reach second point from first point? Why its movement between those two points wasn't instantaneous? Why and how it maintained velocity and direction if it had no interaction with the outer world. These questions must be there in a mind of the thinker of physics. Is there any navigation system in the mass? Is there any internal clock inside the mass? Does mass do its work without help of any such things?

In this theory of Brahm it is considered that all these happens, because the vacuum is not a blank, but it is also full of Brahm, and object was never left alone. The interaction between object and Brahm determined its velocity, and its direction of motion.

If Brahm is also there, then how inertia works? Why Brahm don't provide a resistance to motion with constant velocity in a space while we need a force to move something in space!

We need a force to move something in Brahm, does that means that Brahm is giving a resistance against the motion of that object? The answer is both yes as well as no. Brahm provides a resistance to the change in a motion state but not to the motion at all. Actually, no instantaneous displacement is due to resistance by Brahm that don't allow infinite velocity. How!

Let's imagine a mass, globular mass within Aether. As described in relevant chapter that Brahm is a gravitational environment which contain multi-scaled flows. Flow of waves, flow of masses, larger, smaller, very small and smaller. When a globular mass stays stable in Brahm, it has a balance of such flow from every sides. Let's name this flow for conveniency. It may be a wave or mass, but ultimately it carries energy momentum. So, we will call it, flow of momentum carriers. We have used the word Flux for these momentum carriers in rest part of the book. Flux or tides within Aether are momentum or energy carriers. In case of standstill body within Aether, there are equal tides from all sides. When push is applied to that body with another object, a momentum transfer occurs. Only moving thing can move another object. So, what is a moving thing inside Brahm? It has asymmetrically distributed tides, which surrounds primary mass units. In a case of the moving body, there are tides dragging all unit masses of that body. Those tides flow that body in a particular direction. Not all extra tide is transferred, but as per newton's third law, part of it is reflected back, that balances the object which moved that body, which had imbalanced surrounding tides before the event. So, imbalance of FMC, "a tide" is transferred to stand still object which starts moving due to that tide. If mover object transfers all extra tide, then it has a balanced tides on all sides and it will be a non-moving object in relation to Aether then. So, aethereal tide is a reason behind inertia. Resistance felt to change the balance of such tide because during tide transfer summation of tides of each object occurs which provides back thrust as per the Newtons third law of motion. Once a high tide created over the mass, it continues to move the object. When obstruction stops such object, then tide may damage the structure of the object by trying to crush it. If an object has a sufficient strong structure, then object bounces back. Objects are made of compound waves and tides are unidirectional waves or moving compound waves within Aether. Wave dynamics plays role here. Unlike waves on the water surface. Here tides or compound waves don't disperse in medium and lose their identity because all waves are stabilized by fine flux within Aether. So, here waves are stabilized waves and not dispersible waves.

This provides an insight that there is a basic difference between moving and standstill object inside Aether, moving relative to Aether. But we don't know the technique to identify these tides. These tides may have effect on some property of the object. This may affect vibrational property of the object also. gravity is a shadow in a field of pushing flux, so gravity creates imbalance in tides within its shadow, so it may change property of the atom or subatomic particles of the object being in that shadow.

Inertia

When we push something then actually a mechanism creates a mechanical arrangement that provides a flow of tides in relation to the object. That will create a tide for each mass units of that object and thus transmit the momentum to the object. That tide is actually aethereal waves and thus it drags mass within Aether because mass is also a cluster of waves within Aether.

Actually, tide becomes part of the mass, and it increases momentum and also mass of the object. Mass has two components of wave motion; one is circular or orbiting and other is linear. Tide unites with waves of mass unit in a way that it drags wave complex in a direction of the motion of tides. When such two similar objects having motion in relation to Aether collides with each other, that would not cause a loss of the momentum. Momentum of the tide units play its role. On collision, that stays with parts of the structure, and it does the following things. Disrupts or crushes the structure of the material of the object by providing pressure at every point of the object, ruptures some parts of that object that radiates as particles and waves, that produces vibration at atomic level, that produces heat, that causes gross vibrations that produces sound, etc. So, tide does it work and stays connected to mass units. Those tides may not change the morphology of the mass units. This is because when tides unite with spinning primary as a wave cluster that creates imbalance on the two side of the primary mass and mass moves in Aether in response to this imbalance which don't low the imbalanced tide to change the shape of the primary mass to much extent. Spin of the primary mass have a centrifugal force that also resist change of shape of primary masses. Thus, tides once unite with masses it becomes a part of the masses.
So, I think inertia is explained.

Time and tide wait for no-one.

This explanation shows a non-resisting nature of Brahm and continuously acting nature of the tide. We don't see here that an extra tide strikes a body and get reflected back and leaving no continuous effect on the momentum of the body. What we see here is, a one extra tide continuously acting on body until it will be balanced by another tide from opposite side. When a mass pushes another mass, that is an aethereal tide which get transferred and not a mass transfer, still it is a momentum transfer and also adds to the mass of object receiving tides. If we see that tide as a curve in Aether, then also it could be explainable. So, motion is result of an imbalance of tides or pushing force on a body in an Aether. Such imbalance can be sustained only when there is a no obstruction. Suppose if a moving body with such tide, when stopped by another bigger heavy mass, then such tide will get transferred to that mass and moving body will have resulting balance of tides. This means that if a 1 kg object strikes with a steel wall embedded in a rock of mountain, then also that 1kg object transfers momentum to that wall. Because of large size of object assembly resultant effect remains undetected.

Now we have some clearer picture about an object in a motion and objects in relative motions. Motion of the object tries to maintain an object in a state that tides of push gravity from all sides remain balanced. When imbalance occurs, free object moves in a space and that motion increases relative value of opposing push gravity or tide, and object gains a balanced state. Shape of the primary masses don't change due to tides that moves mass in the space because even there is an imbalance of tides surrounding mass, a motion of the mass in response to that imbalance, balances the pressure from both sides and thus maintains the shape.

7. Time

It is now clear that time is not subjective relativistic fourth dimension but still can be considered as a fourth dimension in relation to Aether. There is a no relation of time to the movement of the objects relative to one another. So, here definition of the time will be in relation to Aether. As shown in the chapter "Honi", Time is there because events are there. Events are there because the existence in the universe has motions, they have a state. The smallest bit of time passes in general when there is a smallest difference in two consecutive 3d map at anywhere in the existence. What that could be? In aethereal world at Aether scale, a smallest full movement happens when one unit of Aether passes its motion to the nearby next unit. This time is at the Scale of Aether in concern. Even if there is a no perceptible movement at Aether scale, still General time is passing because units of Aether are still in a place which suggest that there is a moving thing called energy flux, which is stabilizing Aether structure. So, according to this theory Time can't stop generally. Why am I saying that even if there is a no change in a structure at all, even then time is passing for it? This is because object has a relation with the world. Even if no change is there in an object, even then there is a change in a world related to that object, and object is no longer standstill in a standstill world.

Absolute or General time:

If a selected area doesn't show a change in the two consecutive frames, it doesn't mean absolute or General time is stopped, a change anywhere else shows a moving General time. Change is the rule which is continuous, and which cannot have a pause. If absolute pause is possible then world is impossible because only motion can create a motion or motions, so if all motions are paused then there could not be a trigger that can start motion again which may spread throughout the Nature. But it may be possible that all motion can get trapped to some extent in a selected volume and remaining portions of the volume could acquire no vortices of waves due to absence of aethereal waves and leads to just vibration of aethereal units which is no material state at Aether scale. This state can explain a state of one of the many places before the big bang in an isolated volume area. Spread of motion from that localised sequestration to elsewhere in that volume or conversion of random vibration to aethereal waves can be considered as a material forming theory as described in "Vishnu Purana". There is a no big bang here but there is a slow material or mass formation which can lead to physical world. That state when only minimal motion is there in a volume area is called a "Neerav" state. Which is a pure peace and having no effective motion, no event, no matter, no wave and no local time. But still, it has a life span because time is still passing in a sequestered area somewhere at that scale and also time is passing at a lower scale. So, absolute time is not ceased which can never get ceased. There could not be an area which could have a no attachment with surrounding. Everything always stays connected and absolute isolation is impossible. Absolute blank is also an impossible thing. Absolute singularity is also impossible. In case of Nirav state, time is standstill at aethereal scale, but still aethereal units exists, means it is supported by push gravity from surrounding, means there is a no ceased motion or ceased time. If all motion ceased, nothing could be there as imaginable within the framework of this theory.

Local time:

Smallest Change in the arrangements of 3d map of existing things in a two consecutive frame shows a passing of local time with motion of smallest units at the scale. If volume outside the selected frame is showing a change in a time, but a selected frame of volume doesn't show a change in 3d map, then it can be said that local time is standstill in a selected volume. There may be various factors that can affect rate of the local times of various volumes. This local time is not a concept like a local time in the special relativity, because this don't have any relation with relative rectilinear motion between two frames. This local time is just a measurement of change in arrangement of mass in 3d map of selected volume. But this idea has restricted practical value, because we can't know the exact arrangement of mass in selected volume by any means, because there cannot be a lower limit of smallness. So, we can never say that there is a no changes in the 3d arrangement of mass in a selected volume. But we can restrict observation to some scale to talk about scale specific local time. We can categorize time based on levels and scales. Time at atomic level, time at subatomic level time at further fine levels or time at different aethereal scales. When we freeze the motions at some levels we can slowdown processes that occurs due to motions at that level, for example freezing a thing by decreasing its temperature can slowdown processes at molecular level powered by heat of the thing, but we can't achieve such slowdown state at sub atomic level, because circular motion of the vortices involved at a sub atomic level are essential part of the existence of the matter, change in that can disrupt the existence of the matter, so slowing down of the time near standstill, when motion of the vortices in Aether are involved is logically not possible along with those masses remain preserved. A motion can't be stopped at all. It is impossible to eliminate motion of any moving thing. Motion can be transferred from one carrier to another but ultimately motion sustains its existence. Existence of the matter and that's why the existence of Aether is due to complex vortices. Vortices carry continuous spinning motions. These motions are the existence that could never be eliminated, so, time just run as the motion and displacement is happening continuously, and time can never stand still in any existence. Aether has a fixed rate of energy transfer; at that rate it transfers a motion as a wave. This rate decides the speed of events in Aether. If we consider a finite maximum speed of wave in Aether, then we can simply calculate that a linear wave has a maximum speed, and a vortex has waves moving in orbits, and their circular velocity must affect its maximum linear speed. So, non-moving vortex in Aether have a maximum circular spinning rate, while moving vortex has a less circular velocity or less spinning rate. It would be equal to maximum possible orbital speed minus linear velocity. If time is dependent on the spin rate, then non-moving vortex should experience maximum time rate, while moving vortex should experience less time rate, or more rate of events or changes in unit time. This seems similar to special relativity, but it is not there. Here there is a reference frame of Aether and no relative motions of objects with one another are considered. So, here also fast-moving object shows slow passage of time, but motion is in relation to preferred reference frame which is universal reference frame Aether. Think about one vortex made up of infinite small vortices within it, if we move this vortex in one direction with constant velocity, does it change the circular motions in small vortices forming that big vortex? We are talking about change in the rate of local time and not a relative time. Relative time concerned with change in the perception of time when two objects move in relation to each-other and not in relation to

Aether according to Einstein's Relativity theory. Acceptance of Aether is contradictory to the postulates of the Special Relativity. The special theory of relativity doesn't fit in the framework of this Aether-Brahm theory. So, Time dilation works here, and it is according to the motion of the units of an object in relation to Aether.

Some fundamental questions related to time could never be answerable. For example, when was the very first beginning or inception of time? The first ever origin of the time remains fundamentally unimaginable. Was there any existence before the prime inception of time? Could there be anything like a beginning of the time? Even could that be possible in a logical imagination! Time has a value depending on the scales which is described in chapter scales.

Singularity

Is there anything like singularity of modern physics can exist within the scope of this theory?

This theory rejects the relativistic blackholes. Relativistic Blackholes can harbour singularity, but they are not here to sustain it. So, no such thing is possible here. It is described above that there could not be a place with no moving things. Where there is a movement there is a time and where there is a movement and time there can't be a singularity. Centre of the galaxy is a relatively gravity free area, but that can't be considered as an area having singularity. We know very little about it.

Topic like a time dilation is explained in the chapter 19.

8. Light

History of Light in science:

Evolution of understanding about light in science:

- Scientists like Isaac Newton and Christiaan Huygens debated the nature of light, with Newton proposing a particle theory and Huygens a wave theory.
- In 19th century, James Clerk Maxwell's electromagnetic theory demonstrated that light is an electromagnetic wave.
- In 20th century, quantum physics brought the understanding that light has a dual nature, behaving as both a wave and a particle (photon).

Light and Aether hypothesis:

In 19th century understanding of the nature of waves and nature of the light as a wave gave rise to the concept of "the luminiferous Aether". When it was established that light behaves as a wave, a belief in a medium which should carry this wave imagined as a Luminiferous Aether. They observed that all kind of waves required a medium to propagate through. Sound in air and tidal waves on water surface are the examples of waves propagating through mediums. Therefore, it was hypothesized that a pervasive, invisible substance called the "Luminiferous Aether" filled all of space, providing the medium through which light waves travelled.

Challenges and refutation of the Luminiferous Aether hypotheses:

As described in the chapter 2 of the first part of this book, Motion in Aether and its drag effect was the priority in the end of 19th century in the field of experimental physics. Logic was applied for the existence of Luminiferous Aether, If earth is moving in relation to the sun, and if Aether medium is there, then it can't remain stationary to the earth during whole year.

So, if Luminiferous Aether medium is moving in relation to the earth then any wave which is created within vicinity of the earth must have drag effect by relatively moving Aether.

Aether was hypothesized in 3 possibilities.
1. Stationary in the space and moving in relation to the earth.
2. Stationary in relation to the earth means a dragged Aether.
3. Partially dragged Aether.

Aberration of the light phenomena ruled out dragged or partially dragged aether. Now there was only one possibility remained and that was a stationary Aether in a space which was moving in relation to the earth. The famous Michelson-Morley experiment in 1887 aimed to detect the Earth's motion through this supposed aether "Wind." The experiment yielded null results. It found no evidence of Aether wind as so of the stationary luminiferous aether also. Along with the Einstein's special theory of relativity this result ultimately lead to abandonment of Aether concept. In present time it is believed that light is an electromagnetic wave, and it requires no medium.

Let's think again about the result of the failed Michelson- Morley Experiment's result:

We can assume behaviour of wave in Aether in two forms.
1. It has a constant speed relative to Aether.
2. It has constant speed relative to source.

If the speed of the wave (for convenience, we will take light in further description.) will depend on the source of the light, it is called emission theory. But findings of the observation of the binary star system contradicts with this theory, so emission theory is not accepted.

These observations support the finding that, once the light enters space it acquires a finite speed irrespective of the speed of the source of the light in relation to observer. So, star light from the star moving away from us, or moving toward us takes the same time to reach us. But from our place we can observe only light which is coming in our direction. We can say that a "to & fro" movement of source relative to us doesn't change the speed of light in space. But what about side-by-side movements? We cannot observe it directly. Suppose we are at point O in the figure below.

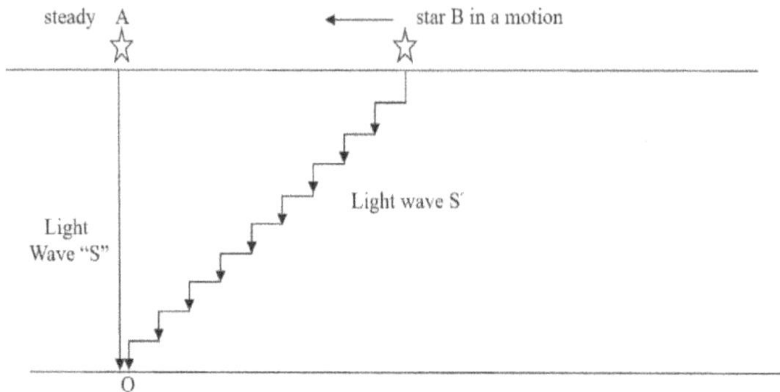

Figure: 8.1

Light from star A will travel less, a minimum straight distance from us, and it will reach us in a specific time.

Star B is moving perpendicular to the line joining star A and us. Imagine two light waves from Star-B. One is emitted in a direction facing directly to us, another is emitted in a direction parallel to light wave "S".

Which one will reach us?

According to emission theory, the first one should reach us. But if we look at the Michelson Morley experiment, the answer is different, the wave parallel to light wave "S" should reach us, this proves emission theory partially true. Because light wave S' travels a longer distance in a less time. But science has observations that don't agree with emission theory. So, for emission theory to remain wrong there must be a first wave which is facing toward us should reach us.

So, what I am explaining is two component of light propagation. The speed in the direction of the emission remains constant, but the movement of the source in perpendicular direction of the axis of the emission line affects light propagation. Light gets dragged while propagating in that direction also as shown in the figure.

This was the explanation that remains valid for the light as a wave or as a particle.

Here It is a clear view that light can't travel without medium if it is a wave, and light can't exist without gravitational environment if it is a mass or particle.

Wave nature of a light as a particle can be explained by helical path of moving charge in magnetic field. There would be continuous field interaction of light particle with Brahm leading to helical motion and thus results in a wave nature of light.

So, here I am not choosing a nature of light as wave, or particle. This is a matter of experimental verification. But all these two can exist within the scope of this theory of Brahm.

Light as a wave can be categorized as aethereal wave. Aethereal waves are fastest thing in Aether to be propagated. When waves form vortex, it behaves as a mass, but aethereal wave don't behave as a mass, and it don't show the property of time. This doesn't mean that there is a no ageing of aethereal waves. Tapering in Brahm leads to gradually tapered waves which changes their property of frequency and wavelength, and this phenomenon is perceivable as astronomical red shift. Tapering of light is explained in a topic energy waves in a chapter 6 in this same part of the book. This phenomenon also restricts the span of the visible light during its journey. This shows that light can't remain as a visible light after specific distance in a space in Aether. Light gets converted to invisible wave after some distance with long wavelength and low frequency. What could be the next form of light ahead after a form of radio waves is a matter of predictions.

Limited visible or observable universe:

Observable universe is just a part we can observe with Light or other EM waves using telescopes. This limit is believed to be due to Big-bang and expanding universe. But if we inquire this in a perspective of above description about the nature of Light, we can understand that this limited volume of visible universe is there because light faces tapering and can remain visible after certain distance. If we move toward any direction, a boundary of observable universe would also move with us. This is not detectable from a motion within our solar system because the distance light travels as a visible light is a very large. So, according to this logical explanation a boundary of the observable universe is not a proof of the Big-Bang.

Relation of light with electricity:

As explained in the topic c electricity, light affects the spin of tori. Thus, light can form an electricity if it is properly received by a medium containing tori in proper orientation and proper situation. Electricity can also affect the spin of the tori and a deviation angle in the circuit can extract a flow of light passing through aligned series of tori. So, light can create an electricity and electricity carries light.

9. Controversy in Big-bang and Expanding Universe theories of the modern science.

Not going deep into the history of big bang theory and expanding universe theory, but at present in mainstream physics this concept is the core of the model of the universe and standard model of the physics.

There are two main reasons behind this belief.
1. Cosmic microwave background.
2. Red shift of the light coming from the far object.

1. Cosmic microwave background

The Cosmic microwave background (CMB) is there. In 1964, Arno Penzias and Robert Wilson, working at Bell Telephone Laboratories in Holmdel, New Jersey, were conducting experiments with a sensitive radio antenna. They encountered a persistent unexplained background noise that interfered with their measurements. That noise has no particular source, but it was across the whole sky.

When the radio telescope was placed toward sky it was found that a certain amount of the radio wave was coming from all directions. These radio waves were coming from all directions and going towards all directions, with the same amount of energy within them. These make the background temperature of the space more than absolute zero. CMB was predicted by some physicists as an afterglow of the big bang. Later that noise was considered as a predicted radiation. This was considered as a strong evidence of the Big-bang.

Penzias and Wilson were awarded the 1978 Nobel Prize in Physics for their discovery.

CBM is considered as a "afterglow" of the Big-bang. It is made up of faint microwave radiation that fills an entire observable universe. It is believed that it originated during Big-bang, and gradual cooling of the universe have left such an amount of microwave radiation background. CMB is a form of electromagnetic radiation. It has a temperature of 2.73 Kelvin.

With insights from this theory of Brahm, the claim about the cosmic microwave background is, it is an expressed portion of energy flux moving in all direction in Brahm. In other words, it is the expressed portion of the turbulence in Aether. As described in this book, Aether can't be at a calm state, component of Aether can't be in a stand still situation. Brahm has much more energy moving in all directions, which expresses itself as a push gravity, as a light waves, as other types of waves, as a magnetic field and as particles. Microwave background is one of them. It is described in the topic below that a microwave could be a tapered or transformed light originating far away from observer. It is the intermediate state of light during its whole life span. Gravity forming waves may be the next fate of it. So, in this theory it is not a thing that has a common origin or same time of origin, it is a part of continuous process happening in the universe. It shows a vibrant hot state of the always existing Brahm. Light has a finite life span as a visible light. Due to wave transformation in Brahm light loses its character as a visible light progressively and becomes microwave during its journey. This is the reason why there is a limit of the visible universe. The universe exists even beyond that limit, but due to the short life of light we are unable to see beyond.

2. Red shift of the light coming from the far object:

In one of the few theories of modern physics, it is believed that the universe is formed by big-bang and the universe is expanding like an inflating balloon. So, every object in the space of the universe is moving apart from each other. But that expansion doesn't affect the size of the structures like atoms, particles, stars and galaxies. This expansion occurs just in a way that it only increases the distance between this structures. In short there is an expansion of space but not of material things. According to modern physics, Distant structures like galaxies are moving away from us at enormous speed, with speed more than that of light. So, light emitted by distant galaxies shows a doppler effect and a redshift which can be confirmed by absorption spectrum of elements in star lights. When a wave is generated by an object moving away from the observer, an apparent frequency becomes lower and apparent wavelength becomes longer. This is called the astronomical redshift. All sufficiently distant light sources show astronomical redshift which corresponds to their distance from the earth and so corresponding to their speedy motion away from the earth. This observation is known as Hubble's Law that implies that the universe is expanding. The value of red shift is denoted by alphabet "z."

Some observations like quasars just near to galaxies provide some weired observational data, which don't fit within this expanding universe model. But science keeps ignoring such observations and keep holding their beliefs. The structures which are at same distance from the earth are showing different z values of redshift suggests that the reason behind this redshift like phenomenon is somewhat different. An American astronomer Halton Arp worked on such observations and found controversial data in observation of quasar and galaxy pairs. He found more than hundreds of such observations that challenges Hubble's Law. Halton Arp's research focused on peculiar galaxies with unusual shapes and structures. He argued that quasars are ejected from the nuclei of these galaxies and their higher redshifts are not due to their distance from the earth but that is their intrinsic property. So, alternate explanation for this astronomical red-shift and cosmic microwave background is very old demand of time.

Another controversy arises from this belief is that, according to this expanding universe and big bang theory, if we calculate a centre of the big bang, then it becomes ourselves. Like a postulate of SR containing observer specific constancy of the speed of light, this is observer specific centre of the big bang. In respect to the observations, Blast must have occurred at the place from where we are observing, because microwave background shows same intensity from all sides and distant objects shows same z values of redshift in all directions. This seems weired part of the big bang theory.

Other explanations exist that can explain this red shifts as a tired light, but that type of effect may not be distinguishable from the astronomical red shifts. Tired-light effect is also an option, but which is not widely accepted by modern science because scientists don't like that idea, because that doesn't fit in with the expanding universe and big bang model of modern science. Recent advancement after considering findings of JWST, scientists are now feeling uncomfortable with the old model and they have introduced advanced model with acceptance of both, redshift, and tired light phenomena together. But they have not completely denied the expansion of the universe. They just made only necessary changes in the old standard model to fit the new findings of JWST.

Wave transformation phenomena explained below is the explanation for such redshift-like effect according to this theory of Brahm.

Wave transformation or Wave tapering phenomenon

This is explained in a topic energy waves in chapter 6 of this same part of the book. When there is a continuous compressing tapering force or pressing force acting upon the wave or on an object in a wave motion, then that force makes changes in the frequency, wavelength, and shape of the waves progressively, this phenomena is called wave transformation phenomena or tapering of the waves.

The same thing occurs with the light waves in the space where a pressing or tapering force due to elasticity of Aether or effect like a push gravity, light wave gets tapered over a time and it becomes narrower gradually, that looks like a decrease in the frequency and increase in the wavelength. This transformation mimics red shifts to some extent and is mistaken for Doppler effect which contributes to the foundation of the Big-bang and expanding universe theories.

Figure showing such an effect.

Figure 9.1

This is just to explain the phenomena, actually it takes many more cycles to gain such effect, and in case of light-wave, its width decreases and the length of wave increases. The change is not simultaneous with decrease in the energy content of the wave, but before vanishing of the energy, this transformation occurs due to nature of Aether medium propagating that wave.

This is the explanation for apparent astrological redshift. This provides a relation between distance and amount of this effect. Tiredness of light is also applied when light passes through materials. It also depends on the type of light when it originates. Gravitational fields in the path of light also affect the redshift.

It is well known that different data of z value of the astronomical redshift phenomena of distant lights create confusion after recent observations by JWST. This suggests that there is something serious wrong in the present understanding of modern science about the universe.

10. Aether cycles and a steady state.

Aethereal cycles

It is possible that the whole existence including Brahm with Aether could be in a steady state, with local changes but no gross level changes. All scales of Aether remain always there and has a no wear and tear effect. It is also possible that there is a universal amorphous unistructural permanent Aether is there and it is the prime cause of the existence, and everything manifests as waves within that Permanent Aether. But this would be like stopping a thinking beyond some extent about smallness.

Other possibility is cyclic changes in the existence for destruction and recreations. This may be limited to topmost Ather layer of Nature or limited to some layers of Aethers or May involve whole Existence.

Let's apply logic for this option.

Why a cycle in Aether and the creation is there!
Logic is simple, it is there because there is a no other option, and change is the rule due to constant imbalance!
Why?
Let's see,
From Honi we reach to the conclusion that a changing universe could be the only possible possibility that can accommodate real world within itself. Now let's think further that, there is a flow chart as below.
Let assume a state "Z", which is a random state we consider as a Starting of the universe. Let assume a state "Y", which is a state that is a state just prior to "Z".
Other states could be "A", "B", "C" ....
In the chart shown below, states between "C" and "W" could be infinites. "W" always leads to "X" and "X" always leads to "Y".
"Y" always leads to the state "Z" which is considered as a beginning state of the universe here. So, this flowchart shows that when there are infinite possibilities, it always led to cycle.
And that's why there is an aethereal cycle.
This pattern is not compatible with the will power interference.

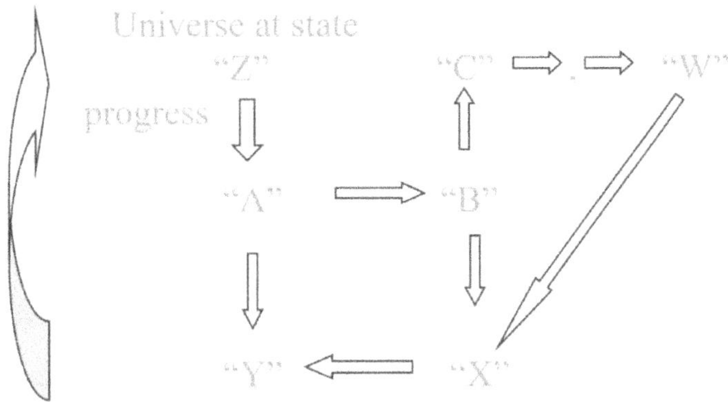

Figure 10.1

Let's think in more descriptive way,
Because there is a flow of the energy waves in Aether which makes it "Brahm."
Two possibilities are there.
1. Aether has a finite amount of energy and that can create vortices within it and creation and demise of vortices are there but no gross level change in aethereal state. World is in a constant steady state with local creation and destruction within it.
2. Gross level cyclic changes are there.

What exists is the Nature. The distribution of something in the Nature is such that it allows a shift of position of something within existed Nature. This allows a motion or a movement of the something. That allows a wave formation and thus allows an energy to exist. This motion may be distributed evenly everywhere or within a limited area and allowing a world within a steady state, but a balance between energy and mass can have a deviation toward energy side or a mass side. More mass and less free energy is there, when more vortices can accommodate much energy, lead to a calm world. Less mass and more free energy are there, when less vortices accommodate less energy, and more linear motions are there which can result in a hot world full of energy and hot environment not favourable for mass formation. There could be a shift between these phases, and in between that there is some interval with perfect balance that can provide a scope for temporary steady world to exist. Such transition from cool calm state to the hot state can be viewed as gradual serial changes or phasic changes. So, there could be cycles in an Aether which can generate a dynamic state of Aether world. State of the universe could be defined by distribution of motions within aethereal units. We can imagine various states accordingly. We can't imagine a size of the universe, we have imagined a limited portion, it is a fixed portion of the universe that can provide a room for enough variation and possibilities that could lead to aethereal cycle. The relation of such portion to higher scale or more huge areas must be there which represent bigger CYCLES at higher scale or in more huge volumes. So, here also size and number of cycles are unimaginable and infinite. There may be a specific pattern and interval for this physical world or Brahmand but there is a no logical way that could calculate them.

Galactic cycle

You can also call it a matter cycle. When energy moves in Brahm, there is an incident where energy waves unite in a way that it forms a vortex which has a property of the mass, and we refer it as a primordial matter "Kan". Such Kans (vortices) formed separately but gathered as a cluster and made a cluster of the Kans called "mass cluster". Again, these mass clusters get pushed toward each other by push gravity and make a cloud of distributed matter. Atom and elements formations occurs due to push and torus magnetic fields. When gravity again works, there will be a formation of large vortex known as galaxy or Quasar, and formation of massive objects like stars within it. They get arranged together and moves toward each other, because of their polar properties and planner movement, they all orbit around their common centre of mass. You can refer to Torus formation topic to understand these phenomena. When a giant vortex named galaxy is formed there is a flow of flux from both polls toward a middle transverse plane. When energy flow unites in this plane from upward and downward sides with an orbital motion to some angle, it tends to fuse with each other, and a further matter formation occurs. These matter formations progressively become intense and during its journey on the circular path in a plane far away and around the centre of the galaxy it gets condensed, and forms proper mass, and it becomes clustered, where push gravity works to form stars and star clusters. This mass becomes stable entity until it reaches a centre of the galaxy where it gets annihilated due to lack of the gravity push because of the shadow of the mass distributed in a galactic plane, energy waves from above and downward interacts with annihilating mass at the centre of the galaxy which is a hole but not black at all with a way above and downward and placed as a hole in a centre of the galaxy. Larger the galaxy, clearer and complete is the annihilation, smaller the galaxy, incomplete and fuzzy annihilation, resulting in a formation of visible energy and material jets.

A spiral type of galaxy once formed behave like a machine that gather primordial matter from surrounding and giving the way toward centre of the galaxy and ends it there. This is a galactic matter cycle.

A matter thus formed if finds its way away from the galaxy, then it will remain stable until it will get its way to the centre of the galaxy somehow. Matter can remain stable in the free space outside of the galaxy because gravity push is not limited to the galaxy alone.

So, galaxies and centre of the galaxies plays role in the formation of matter and its dissolution in Brahm again. This journey of waves as a clustered matter is called "Sangh" means a togetherness. So, matter born by pairing two units of flux, and it has an end by annihilation into separate bunches of energy flux at centre of the galaxy. [This mimic a story of a character named Jarasangh in a Mahabharat, who born in two separate pieces from his two separate mothers then get united to become a one person and dies when he was teared apart in two pieces and thrown away by throwing two separate parts in two different directions. Not just this, but I found so many stories and paintings in Indian Hindu mythology that has close proximity with various portion of this theory. So, I like to preserve some words in this theory as they are in those stories]. The stability of matter depends on its structure plus gravitational push. This gravitational push is not constant at every point of the galaxy. It is higher near periphery and decreases gradually as the shadow of matter from surrounding increases as the distance toward centre becomes less. So, value of gravitational force and strong force also vary according to place in the galaxy. The atoms with higher atomic numbers, which are not unstable at far place, may become unstable if placed nearer to centre of the galaxy and the centre of the galaxy is a reactor where all type of matter becomes unstable and get annihilated into energy. So, galaxies are structures which collects or creates matter and turn it back to energy. So, they are also source of the energy waves, and they are recyclers.

Rethinking on aethereal cycles

There are two ultimate existences, one is Aether and other is energy. Now we can understand that the energy is nothing but a motion in Aether. We know that there are two type of energy flow, one is linear (may be curved at higher scale or at very large distance) and other is curved or circular at normal scale. So, there is a possibility that maximum energy can get converted into circular waves and there will be whirls everywhere which get arranged in such a way that there will be symmetry of that whirls.

If possibilities for Brahm are imagined, two are there. When energy dominates and where matter dominates.

This may be two extremes of the existence, and there could be phases in between them. This may explain the formation of the material world and its annihilation. Use your imagination for this. What I would like to believe is, there is a steady state in the universe and the creation and annihilation is limited to galaxy levels. But as we think higher, we can understand that our brahmand is a cell of higher universe, and it may reside somewhere in the higher galaxy, so there will be a time when our brahmand will also have to be annihilated when it will reach to centre of that galaxy of the higher universe. It is also a possibility that Our Brahmand is a primary gravity push forming corpuscle of the higher scale or just a unit of Aether at higher scale and it wouldn't be destroyed at the centre of the galaxy of the higher scale.

There is no higher limit and there is no smaller limit.
There can be infinite universe within a primordial matter, and there can be infinite universe at higher scales. We can't know that up to what scale a galactic centre can annihilate the matter, and beyond what scale matter remains unaffected even at galactic centre. But we can assume that if corpuscles are there which are contributing to energy flux which stabilizes aethereal units then we can say that there could not be an annihilation beyond that.

Aether matrix is like a hexagonal close-pack symmetry, we can see such pattern in a 2d, on the back shell of the tortoises. The cell of Aether are stables because there is a lower scale Aether within Brahm is there which provides energy flux as waves in that Aether and thus provide a push gravity to the cells of this Aether. If it would not be so, then these Aether's cells must disintegrate. Now if we can think that for that lower scale to remain stable, there must be an effective push gravity because of Brahm of further lower scale. So, for each scale there is a need for lower scale, So, in and in, within it there is an infinite sequence of finer Aethers. If we consider a normal scale as one tortoise, we can say that, that tortoise supports our universe, then we can say that a smaller tortoises support these tortoises. Further smaller and smaller tortoises support tortoises of higher scales and get support from lower scale tortoises.

So, on and on.......in and in, within it.

We can't get the rid of the infinity.

Neither at lower scale, nor at the higher scale.

Mystery remains mystery.

We know something, but in comparison to total existence, approximately we know a little. But if we consider Aether as an elemental substance and rest as a waves and vortices of waves within it, then maths of infinity is not linear here, but Aether provides a band within this infinite scales. The sub world "ther" in a word aether means "a bunch of layers" in a language Gujarati. This also co-relates with its nature.

11. Bhav

This chapter is just here to apply some terminology and nomenclature to the parts of natural massive objects.

Bhav (to be) is a term that represents anything which exists and have separate identity from Aether, and Sambhav means a possible existence.

Bhav (existence with expressions more than just Aether) has two portions,

Self and AURA or Halos.

Self is a thing proper, and Aura is an effect of that thing in Aether.

"Swabha" is self-proper, and "Prabha" and "Abha" are effects of Swabha in Aether.

ABHA, PRABHA & SWABHA

In relation to materialistic existence, we can divide portion of the structure as own self, affected aethereal flux and radiation turbulence.

Swabha (Swayam-Bhav)

When we are talking about the torus, a spinning disc or spheroid is the own self or swabha, which is real material thing. But if we look inside this swabha, its forming units includes affected aethereal flux also which is part of smaller tori forming material of this Large Torus. Swabha is the main material part that applies resistance effect within Aether that affects aethereal flux and forms a Torus with its field. Relatively size of swabha is smaller than total size of the torus including Prabha or magnetic field when torus is formed enough.

When there is a one vortex or torus, portion of Prabha around total mass is bigger. When there is a cluster of such tori, portion of Prabha around the total mass becomes comparatively less. For a very big cluster total size of cluster is far larger than a peripheral margin of the Prabha when the cluster itself is not forming a larger torus or not forming a magnet. This provides a sense of thinking about reach of Prabha.

Prabha (field)

Two components:

1. Magnetic component

In the Torus, bending of the Energy flux forms an intertwining winding stream, we call it a magnetic field, which is considered as Prabha here. Prabha has direction and shape also, that depends on the rate and direction of the spin of disc or sphere forming Swabha. It means Prabha is the effect of Swabha on aethereal flux. That is a curved flux. So, Prabha is a part of the outer environment which gets affected by main material called swabha. Flux pushed toward Swabha because gravity push also works on the flux and it bends the flux toward Swabha which causes a field formation. This also indicates that there is even a finer stream of fine-flux that indeed creates a push on the gravity forming flux itself. This also supports the scaled property of Brahm.

Unification with summation of the Prabha can be there, which we can relate with the summation of magnetic fields of two nearby touching magnates. Two small magnetic field forms a larger magnetic field which's size is larger than each of magnetic fields. This explains a potentiating effect of magnetic field summation, which helps in explanation of magnetic fields of spinning planets or spinning stars or galaxies.

Prabha is the area where repulsive and attractive magnetic force works. At Place away from the poles, such effect decreases in an abrupt manner after some distance from pick concentration area, so it is not distributed as a linear graph but as a bell shape with a narrow bell on normal distribution curve if plotted for strength / distance. So, magnetic forces have a less value at far distance in comparison to push gravity effect.

2. Push gravity component

Push gravity component formed due to resisting nature of the mass that provide resistance on the way of waves and wave clusters in Aether. There is comparatively a more distant reach of such halos when compared to magnetic component of the Prabha. This Prabha expressed as a gravitational field of the object in general. This effect has a long reach because it is a negative one. This is due to absence of energy flux due to obliteration by mass, which otherwise must have very far reach.

So, Swabha is a self. Prabha is an affected flux forming a field.

Abha

A field generated by radiation from the object is considered here as the Abha. If we talk about the reach of aethereal waves generated by aethereal objects than it could be termed as visibility of the object. So, Abha is the area of the universe which could be illuminated or occupied by radiation generated from the object. We can also consider this halo as an additional weak Prabha also.

Shobha

Prabha of the object attracts and places some other objects around it, which is not a part of the object itself, but that stays with the object. Such things are considered here as a Shobha. If electrons are considered as particles spinning around nucleus due to its field, then it could be considered as a Shobha also. Shobha has its own effect and potential for interactions. It also generates a field. But if electrons are shooting out and reuniting back to the nucleus and they are also part of the nucleons then they should be considered as a part of Swabha.

Attracted object which is not generated by self could also be considered as a Shobha. Planets of the solarium is an example of such things.

12. Scales In aethereal symmetry

Aethereal Units define the scales.

In 1899, German physicist Max Planck proposed a universal set of units which are known as Planck's units. Plank units are system of units of measurement defined exclusively in term of four physical constants c, G, $\hbar$ and k_B. They are system of natural units. This units have a root in nature. Plank units are relevant in research on unified theories such as quantum gravity. So, different Planck units have inter-relations and relation with constants of the natural things also, like speed of light, smallest possible blackhole, smallest possibly measurable time in a modern physics. So, actually it is not about the smallest things, but it is about units in relation to natural constants.

In modern quantum physics Planck scale is a well-known term. Plank scale refers to quantities of space, time, energy and other units that are similar in magnitudes. That defines very small length at which direct observation is not possible, and prediction of standard model and relativity are not applicable at plank scale. It is also considered as a universal limit of smallness beyond which Law of modern physics don't make a sense. Actually, there is a no tool available that can collect information beyond the Planck's scale. Speculations of scientists are matter of faith. If we talk about the Planck length, then it is a 1.616255×10^{-35} meter or 10^{-20} times the size of a proton. It is also imagined as a size of the smallest black hole in a theory of the quantum gravity which have a no possible observable support. Speed of light is considered as a one Planck length per Planck time. These are imaginations and not real measurable things in a modern quantum physics field. So, Planck level is a limit of measurement by tool available in a modern physics. Subatomic level, atomic level, molecular level, gross level, astronomical level are some examples of levels based on length in modern physics. Though plank scale is defined as a scale in modern physics but in this theory of Brahm-Aether I want to use the term scale for Aether scales only.

In modern physics scientist has observed the max limits in relation to size, mass and length. It is observed that all protons have a same mass, all neutrons have a same mass, all electrons have a same charge. Gravity has a universal constant. Planck constant "$\hbar$" has a fixed universal value and that's why it is a constant. Modern science doesn't know what determines these limits. Why a bigger proton is not possible. Why electrons are not available in different sizes and in different variable charges. Why quantization is there, why uniform pockets are there? There is a no explanation for these in modern physics. If there are limits at smaller scale then why quantity like mass, gravity, size don't have higher limit at astronomical level is also not understandable with theories of the modern physics.

This description was from the scope of the modern physics.

Now let's think in a scope of Aether-Brahm theory.

Why size levels and scales are there?

Reason:

Push gravity is a thing that distributes resisting materials formed in Aether in separate pockets. This is because obliterations by formed clusters restrict the reach of pushing flux toward the centre of cluster after gathering enough mass. This phenomenon limits the strength of the push gravity to a specific limit that cannot form clusters bigger than optimum size. At a place away from the shadow of this cluster push gravity can work normally. These forms separated pockets of clusters. Thus, there is a well distributed clusters at some specific intervals, which forms larger cluster. There bigger clusters can also be distributed at certain intervals of space and further forms bigger cluster. Torus formation doesn't allow all distributions in a globular form, and this leads to variety of shapes and types of clusters formed by push gravity. All these can form specific sized structures with maximum size limits at particular level defined by sizes of the cluster forming units.

So, reason behind quantum nature of mass is the limit of push gravity. Push gravity forms clusters of material. When more and more mass is added, it reaches to a limit beyond what further mass cannot be added to that cluster. This happens in both cases, in formation of the spherical mass and in case of the torus also. Similar clusters can exist in Aether only at the minimum distance required to get enough support of push gravity not affected too much by the shadow of the concerned cluster. Not just a limitation of the push gravity but a torus field also plays a role as a repulsive field that also pushes structures with magnetic fields away. This explains a distribution of the stars in the galaxy also. so, basically limit of the push gravity along with magnetic fields defines distributions of mass in pockets and not forming a limitless single mega cluster without space within it.

Limit of Gravity is based on the density of material also, if material is more porous, it forms a larger and less dense cluster. if material is less porous, it forms a smaller, hotter and denser cluster.

Primary mass is a torus which can form globular clusters. Such globes due to spin may become polar objects. combination of such mixed cluster forms a globe, or torus depends on various factors. Whatever is formed either globe or torus, they have maximum size limits.

Many tori can stay at distance from each other, and they can arrange themselves according to fields they created. That forms a cluster also which may form a stable globe which could be served as units to form Aether of the higher scale. If I talk about scale around us, in relation to matter and cosmos unit of Aether is a smallest non perceivable stable corpuscle or globe which forms lattice of Aether at this scale that allows a waves formation of the width of minimum size which must be many times the size of that units. These waves form primary masses which can form mass clusters and bigger structure at this scale. If we assume electrons or protons as smallest particles at this aethereal scale, then size of the unit of Aether at this scale would be extremely smaller than the size of the electrons. Brahmand a globe containing infinite cluster of Galaxies within itself would become unit of the higher scaled Aether. So, we are assuming Brahmand as units and not galaxies as units. This is just an assumption because there is a no way to calculate the exact distribution of masses between scales. But what we accept is a very stable structure as a unit of the Aether which could be a Brahmand but not a galaxy itself. These are aethereal scales and there are infinite scales of Aether. Aether is such a multi scaled Matrix as explained in a chapter Honi.

Waves in Aether are formed by serial deviation of aethereal units. Those waves when forms cluster that is considered as primary mass. Mass units create spheres. Sphere creates tori and tori further forms a globe. There are multiple combination and cluster possible with these structures. Which leads to giant structures. That forms a stable unit. That may form a higher scaled Aether with different time scale. How two or more waves can form a wave cluster, or a wave couple is easy to understand and thus not explained here.

So, aethereal scales are formed with stable structures which behaves as aethereal units. Levels are described on the basis of sizes of different structures between two consecutive aethereal scales.

Time perception is different at scales.

If we think in relation to aethereal units, there is a close pack symmetry of spheroids or tori, and which forms a 3d mesh. It has units. Unit has size limits. That units have limits in relation to movement in response to forces. It has a limits or spectrum in relation to oscillation frequencies and width of movements. That defines quantization of wavelength and frequencies of aethereal waves also. It determines a speed of aethereal waves also within specific range.

Time depends on the motions. If Aether waves forms toroid mass, then circular motion of that toroid mass units defines time. When a mass moves in relation to Aether still time passes due to spin of units of mass. Sum of both motions, spin and linear motion of units of mass becomes millions of times of length of its size, meanwhile motion of mass units at higher scale can move aethereal units for just a very small fraction of unit length of that scale. So, in higher scale time passes very slowly in relation to time at this scale. From this scale, higher scale appears stand still or very sluggishly moving. During our entire life observation at this scale, we cannot notice any noticeable motion that can define a pattern of motion at higher scale even if technically detection of that motion becomes possible. Similarly, during a fraction of unit time at our scale, the lower scale structures move so faster that we cannot even detect their individual presence. If we imagine a life at a lower scale, than we can say that life of billions of years in their universe get created and destroyed many times in our little second. These are due to scales of Aether. This is because Aether allows maximum speed only relative to scale and size of the structures. So, size scales also lead to time scales.

Relativity in brief in relation to Aether

Einsteins special theory of relativity accepted no universal reference frame and thus no Aether. Dynamic portion of Brahm May fit as a spacetime matrix of special relativity but Aether has no role in special relativity. If we accept Aether, then we have to modify Special relativity as a Relativity in Relation to Brahm.

Every Material thing in a Brahm is formed with waves in Aether. Aethereal waves have finite speed of travels. Couple of clusters of waves acquire circular motion and form a spinning thing which can be considered as a unit mass. Time for that mass depends on its spin. Linear motion of such unit masses acquires with a cost of spins because wave can either travel in a circle or in a line. Sum of linear velocity and speed in a circular motion remains constant due to aethereal property of finite speed of propagation of aethereal waves.

Thus, when unit mass moves in relation to Aether there is a decrease in its rate to complete spin. In other words, time for that mass units slowdowns. This explains Time dilation. If we accept Aether and think of one mass orbiting other heavy non spinning mass, then we can simply understand that an orbiting mass moves with much velocity then a heavy central mass. So, in such pair time dilates for orbiting mass.

This Relativity in relation to Aether and chapter on Light explains results of Michaelson Morely's experiment. Thus, in this theory Time dilation is a result of property of the light in Aether but it is not same as relative time dilation of Special relativity.

If we talk about general relativity, Brahm and Shadow Gravity provides a mechanism of curvature of spacetime around mass and defines gravity as a combine property of mass and Space. As soon as Gravity depends on potential of space and not on only Mass itself, Vicious nature of limitless Gravity no longer exists and things like relativistic Black hole can't exists.

Chapter on Relativity contains some calculations, and it is a lengthy topic. So, it is explained ahead as a chapter 19.

Explanation of Some Observations

1. Quasar jet

As described in the torus formation, when material forming a disc of the torus is insufficient then a gravitational void formation at the centre of the disc remains incomplete, that allows only partial annihilation of the matter and also provides a narrow track for energy waves to pass through. The interaction of energy waves with annihilating matter and passage through narrow track results in a formation of illuminated jets.

According to this theory that jets are not due to any process within a blackhole. This theory denies the existence of relativistic blackhole at the centre of the quasar or galaxy.

2. Luminance at the Galaxy centre

As described in the torus formation, when material forming a disc of the torus is sufficient then a gravitational void formation at the centre of the disc becomes complete, that allows complete annihilation of the matter and also provides a track for energy waves to pass through. The Annihilating matter and passage of energy flux through this gravitational void area creates an illumination at the centre of the galaxy. That is not due to fusion reaction, but that is due to annihilation of matter in the absence of sufficient gravitational push and thus absent sufficient strong force.

3. Bending of the light and gravitational lensing

Explanation 1: Photons may be particles which have a mass. Thus, its path bends in a gravitational field due to attraction by gravity or in other words due to push-gravity.

Explanation 2:

Bent geodesics: Aether around the mass is affected by gravity. Their units show constriction, thus producing a curved path toward mass. As an aethereal waves, light follows the course of the geodesics in Aether to some extent. But this design doesn't allow a fall of light inside the mass in any case. This geodesics are not the same as geodesics in general relativity, because limitless vicious gravity just depending on mass is not here. Here there is a limitation of this chance of geodesic formation because mass is an aethereal wave which may constrict Aether, but mass itself is not a condensed Aether itself. It is a cluster of waves within Aether. What if mass is a condensed Aether? Then there would be a different theory to explain this phenomena or there may be some explanation within the scope of this theory which is not provided here.

4. Rosette pattern eccentric orbit of mercury around sun which is not an ellipse

According to classical physics, Kepler's law and newtons gravity equation, orbit of any planet around the star or sun must be an ellipse. But it is not the case in the case of planet mercury. Such a case is also observed in the case of a star orbiting one centre.

This observation is considered as proof of the time dilation according to general relativity. But this is not the reason alone. This observations are actually proof of the existence of Brahm, and its drag effect. Sun is spinning on its axis. This spinning motion flings the reflected energy flux around it, which causes a drag effect on the mercury when it passes from the nearby orbit area.

In the case of a star, it is something different. Star orbiting a blackhole is not actually orbiting a blackhole. There may be a huge planet or brown dwarf in a pair. There may be a vortex within Aether itself either due to spinning and orbiting mass or due to vortex within Aether material itself. That drags star within a stream and that orbits a centre of the vortex. There is no blackhole attracting the star. Star is in a stream of the whirlpool. If it is aethereal flow or a flow of energy flux or mass in a Brahm is a matter of debate but both of them can provide similar effect.

5. Cosmic microwave background

It is made up of the wavy fluctuations within Aether. It is transformed light. Light doesn't stay as light, as it passes a certain distance and gets transformed or tapered into other types of waves which contributes to the cosmic microwave background. Radiation is radiated by multiple sources everywhere in the universe. It has multiple origins and variations in the time of origin. It is not cooled down radiation due to big bang; it is being formed continuously in various processes. There is a huge stream of gravity forming flux in a space that is the origin of gravitational and magnetic fields. Expressed portion of these flux streams contribute to the microwave background.

6. Astronomical red shift

Wave transformation is explained in the relevant chapter. This results in increasing wavelength and decreasing frequency of light coming from the far objects. Different z values of the light coming from the same object show that it can't be just doppler effect. Recent findings by JWST indicate that there is something seriously wrong in the understanding of the universe by modern physics. Characters of the Galaxies found very far don't match their estimated age. They show features like aged galaxy which doesn't fit in the standard model including big bang theory. This indicates different origins of galaxies instead of from a single source big bang.

7. Michaelson Morley experiment

Light may have two components, longitudinal and transverse. So, when path of the light changed away from the longitudinal axis due to such transverse motion of the source of the light adds its momentum to the light and thus modifying its speed along the transverse component. Necessary experiments have not been done which can rule out such an effect. We can observe this in a Michaelson Morley's experiment where we can see that longitudinal light beam travels more still take the same time to reach the point of interference compared to less distance traveled by other beams of light parallel to the motion of the earth. This shows different speeds of the same light in relation to space.

8. Supernova

In addition to fusion reaction products, star can become unstable at critical maximum allowable mass by push gravity.

9. Hot centre of the earth

That is due to the nature of push gravity and pressure created by push gravity. It is not the historical remnant. Though the earth loses core temperature through conduction, still this doesn't allow a decrease in a core temperature due to continuous process producing heat in the core.

10. Tidal locking of the moon

Tidal locking of the Moon with the Earth is due to magnetic interaction between earth and moon.

Magnetic field of the earth

In all directions the earth has little magnates within it. The magnates oriented towards all directions. Due to earth's motion, magnetic fields which have parallel fields with earth's rotational axis have a summation effect, because of all remains in the one direction while spin of the earth which is not changing. While magnetic fields in other directions have continuous directional changes, and thus they can't unite to form one common magnetic field. Thus, rotation of the large enough mass is a process that may generate magnetic field. Once Magnetic field is generated, it further aligns other magnets and strengthens itself.

13. Calibration of the common sense

This book provides a fundamental base for the logical thinking that creates a sense which generates a common sense and Laws that differs from the common sense used in a modern physics.

1. No stable **empty blank space** could be there around masses.

Empty looking space is neither empty nor filled with a thin gas-like substance. Brahm is a vibrant magneto gravitational expression of substance-Aether-like dense material that doesn't express in a routine, but it can have many interactions with mass and energy waves. That carries energy. That provides inertia. A mass cannot be there alone in an empty space, mass is there due to push from surroundings that stabilize the wave-clusters as mass. If push from surrounding disappear, then mass must annihilate. Mass is not a steady solid; it is full of moving things within it and that moving things form mass. Theoretically there can be an experiment where one can build a large globe of stable mass and put an unstable mass like a radioactive material inside it at a centre then there would be an increase in radiation and instability of the radioactive material. One can also observe less radioactivity of such material in a space away from any strong gravitational field.

2. No miraculous gravitational or other type of **primary pure attraction forces** are there; they are just effects of various pushing or dynamics.

Attraction force is an illusion only resulting from the surrounding push. Matter forming force, strong force, and gravity are push effect at various levels depending on density of masses and distance between the masses involved. Electromagnetic force is due to torus formation and direction of the energy flux in Brahm guided by vortex disc forming a torus with aethereal energy flow. Weak force is there due to limitation of the push force and magnetic separation force, which limits the strength to hold masses together.

3. **No fundamental forces** are there; all forces are results of complex movements of aethereal waves expressed as energy.

4. **Energy** is nothing but a motion of moving things in a Brahm. Motion is one among few fundamentally existed things including something and nothing. It is the relative motion of the wave clusters within Aether that creates and maintain a unit of change called **time**.

5. **Mass** is a wave cluster within Aether that acquires space and interrupts motion of other masses and waves to some extent.

6. **Light** may be a wave like other energy waves, and it don't have any special property. The possibility of being it a mass or particle has not been ruled out by this theory. Light is also subjected to transformation during its journey.

7. Existence of relativistic blackhole at the **centre of the galaxy** is against the concept of this theory, A **void** for normal push gravity at galactic centre is a prediction of this theory where annihilation of normal matter may occur. Before this void there can be an accretion disc but formation of globe due to own gravity, like a relativistic blackhole is not a prediction of this theory.

8. **Maximum limit on a star mass** is placed by limited push gravity for certain density of material.

9. **Maximum limit is applicable over the size** of torus, hollows, clusters so that of planets, stars, galaxies, subatomic particle, nucleus and atom is there, which provides identical sized structures in general with maximum mass. This limit defines quantum scales and max sizes at every **scales**.

10. **Electricity** is not a simple flow of electrons, it is a combined thing made up of spin energy momentum of spinning tori, clusters and flow of energy flux in a trac formed within circuit made by aligned spinning mass units or magnets. Electricity can be stored as a spin of tori; that's how acid battery works.

11. Quantum vacuum is nothing but an expressed portion of Brahm.

12. There is a boundary of our world, and it is the boundary of our Brahmand. Uncountable number of Brahmand must be there in the existence. There cannot be any limit of the Entire Nature. The starting and end of existences like mass, motion and time is beyond the scope of normal logic and beyond the scope of human thinking. No one could ever know the truth of first thing with present limitations.

13. Relative motion is not a base for the time travel, but a motion in relation to Brahm- Aether may be.

14. It is not only energy annihilated by fusion of mass within stars that illuminates stars, contribution of surrounding energy in the formation of radiation energy of stars is more than 99.99%

15. Once a material crosses a certain limit of mass, it can't remain in a solid state, and beyond certain limit it can't remain in a single matter cluster at-all, due to limitation of push gravity. there is a pattern of the distribution of the mass in a gravitational field. Torus formation and stabilization of the mass cluster provides a variation in the distribution of mass in a space.

16. Inertia: Time and tide wait for nothing. The inertia is due to tide within Brahm incorporated with matter itself that defines a moving state of mass within Aether.

17. This book and the science are still not enough to understand everything in general and anything completely. Primary fundamental questions still remain unanswered. Like How was the first beginning. What is the ultimate existence. What is the something as absolute thing. What is infinite. What is the reality! Is the world real, or simulation or something else? What is the difference between absolute nothing and something!

With basic concepts as described in this book, one can expand the boundary of the common sense in relation to the problem or the subject. Mass, charge, electricity, magnetism, light, aethereal energy waves, gravity, forces, all these are explained with mechanical concepts in this theory, that really changes a common understanding about them.

14. Inference

There is a medium everywhere in the world described here as Aether and its dynamics full of flux which along with itself form Brahm. It has property of Multiscale Substance with magnetic and gravitational properties. Energy is moving in this medium from all direction to all direction. This provides a gravitational push everywhere in Brahm. Aether is a matrix within a Brahm. Aether can have waves. A vortex of clustered waves in Aether behaves as a resistance for other flux so it behaves as a mass. A cluster of such vortices forms a proper mass. Such proper mass further can form torus. Cluster of Tori and hollows can form a gross matter. There is an energy flow in modified direction in a torus which interact with field of other toruses or masses, that explains the magnetic effect. Gravitational push has its higher limit, so any cluster at a specific scale have higher limit of its massiveness. This higher limit shows a limit of the attraction force, and it get expressed as a weak force. Huge cluster of neutrons, like a neutron star in a normal push gravity environment is not possible to exists within limitation of this theory or it represents a crushed matter due to pressure of outer matter in a huge mass cluster of a big star. Strong force is also a compound force formed by push gravity and a magnetic interaction. Aethereal waves are waves in the vibrant Brahm. There cannot be a limitless vicious gravity, there cannot be an abnormally large mass cluster in normal push gravity environment. Means neutrons alone can't form a stable cluster. Protons alone can't form a stable cluster. there cannot be a cluster of proton and neutron in very big quantity, means an atomic nucleus has a higher limit, beyond which weak force super wins the strong force due to limitation of the push gravity. Thats why periodic table has a limited number of elements. That's why there are no big atoms of macro size. Larger the mass, less stable it is. A small particle has a halo around it and have relation with Brahm surrounding it, charge is due to spin of the quantum particles.

14. Inference

Everything is an aethereal, every mass is a vortex or cluster of vortices of aethereal waves in Brahm. There is nothing like an absolute solid. Energy is also a motion inside Aether substance, and it is not a separate entity.

There are sure claims and predictions of this theory similar to points enumerated in calibration of the common sense.

This theory brings a sense that allow us to unite science and philosophy. This Theory actually breaks a bridge between philosophy and science and represent a single thing which is the subject of concern in both including philosophy and science.

Theory matches with the Advait-Vad described in Advait Vedanta, an Indian philosophy about the Existence. This raises many questions related to source of the knowledge in ancient time.

15. Residual quarries

Residual fundamental quarries:

This theory doesn't explain everything. Fundamental Questions remains as they are.

Those are as follows.

1. What was the prime beginning of the existence! Or there must be a beginning or not!
2. What is the basic existence, means what is something that is not nothing.
3. What is limit of the smallness? Can there be a limit of smallness possible! Can a math answer this?
4. What about the final boundaries of the Nature. Even if there is a boundary, there must be something beyond it, if nothing is there beyond that boundary than what stabilizes that boundary! If the boundary is not stabilized and it is expanding continuously and what the next to that expanding boundary is a nothing, then up to what distance that nothing could be there? After infinite distance in that nothing there must be something there which doesn't allow that boundary to be tagged as a final boundary. If something is there beyond that boundary than how can we call it a final boundary! So, even in the imagination we are not getting rid of the final boundary.
5. Boundaries of the time, this is also an unsolvable query. What could be the first beginning and what could be the last end. This Is also an impossible imaginations.
6. If we think of the existence of supernatural existences, we cannot calculate from these theory whether there is something beyond these material world or not. Whether they have any interference with this material world or not. We can't deny any hidden existence.
7. Supreme existence: we can't say that Aether is the only supreme existence and there is nothing beyond it which

can carry it, or which can have relation with it. Possibility is there that, aethereal units or primary masses are made up of waves in supreme Aether, which may be a permanent existence, a Parmeshwar (Permanent Aether). There could be something else also beyond our imagination.

8. Simulation: We can't rule out the possibility of the world as a simulation and not a real existence.
9. We can't say if there is any purpose behind this existence or if it is purposeless.

Due to lack of the knowledge of the quantization, exact predictions about Aethereal structures could not be possible.

Remaining general queries due to lack of experimental data or limitation of the technology.

1. What is the size ratio between units of Aether at two consequent scales.
2. Quantitative data of vortices or tori forming a hydrogen nucleus or proton.
3. Explanation of atom, subatomic particles with quantization that could fit with experimental data.
4. Is the light just a wave or a moving particle with halo field, or a fish train in a Brahm?
5. Shape of brahmand, torus like or a hollow like? Or it is just infinitely distributed Nature without boundaries
6. Electron: what are the electrons? What if we compress many electrons together? Can two electrons cancel out their negative charge and become a neutral pair?
    There are many unsolved queries and unexplained observations.

16. My journey to the Aether

This journey is still ongoing. I was a curious child nurtured by my father and I had a habit of asking each and everything to my father whatever puzzled me. This may be because I was a less active, much inert and calm boy. He always answered me in a way that I could grasp the knowledge about the particular topic. We had good conversations about world, life, God, Ishwar, science, tools, animals, stories and many things. It was his teaching that gave me knowledge about globe of earth, gravity, wasteness of the universe, atoms, radiations, power of atom bomb, depth of ocean, limitless size of nature not restricted to the boundary of universe. With knowledge of science, he also introduced "Ishwar", an omni present thing with so many supreme characters like It is free from wear and tear and existed ever without a starting or end. He said the world came into existence because of this Ishwar but he did not involve active interference of the Ishwar in any natural processes. My Father said the world works with laws of nature and there is nothing like a miracle exists. This was a direction for me that if I want to understand the world, I have to understand the science and Laws of nature.

I had never thought of Aether as a medium for everything before my higher secondary education. During primary school days when as a student I came to know about an atom as a fundamental matter unit and subatomic particles as a fundamental units of atoms, Though I was curious, I did not question it anymore and accepted it as it was taught to me as a compulsory lesion of a school syllabus. We were neither used to nor allowed to question any scientifically so-called proven things during that time. Feeds were newer and amazing for us as students, so, what we could do is to learn, memorize and imagine about them. That was an essential primary step to get oriented towards science. I have studied in my regional language so there was no communication gap, and I could get whatever taught to me by competent teachers. They tried to answer some queries, but they knew what should be fed to students at secondary schools. Many times, I have tried in my mind to compare the solar system with atoms. The question arose in my mind about the appearance of electrons and their charge. The question "what is the charge in an electron" was never asked by any student including me. It was explained as a property of attraction toward apposite charge. Electrons and protons were opposite charge carrying things where electrons were negatives and always carrying minus sign above their heads. Though I had basic knowledge during that time regarding planetarium, sun, earth, atmosphere, galaxies, and universe which I acquired by asking my father and watching television. Meanwhile I learned about geography and basic astronomy in school also. I learned some little things about blackhole and Neutron stars also during that days. It was described in the book of the subject Science that there are 10^{11} numbers of stars in one galaxy and same number of the galaxies are there in the universe. Obviously, I rejected that data and raised a question about the authentication of the data. I never accepted that blunder at any age. School education taught me about the source of energy production by sun and other stars, a fusion reaction. I discussed it with my father; he was little aware of it but did not know enough about it. That was a big forte conquered by me. My father was a huge source of information for me in my childhood, then I reached a milestone where other sources surpassed his knowledge in the case of science. He remained a guide for me forever. I did not stop there; I continued to gather information about fundamental physics. I was not interested in memorizing data like radius of the earth, distance of different planets from the sun, numbers of the moons of each planet. I was a good student, and I literally memorized everything given in the science book, but that was for the school exams, and I always scored good. I never went against the exam system or method of teaching. I liked the way I was educated by good teachers. Once a young-man told me "If you want to contribute to science, you have to learn what has been discovered till date". For my curiosity I just gathered conceptual knowledge only that created a science model in my mind with explanation of various things. That model was quite incomplete at the time due to fewer questions in my mind assuming science already held many answers yet to be discovered by me. Physics as a proper subject was introduced in science 11th and 12th STD during 1997 to1999. Only few chapters where there describing cosmos and fundamental physics. Most of the physics was about motion mechanics, electromagnetism, waves and other parts of practical physics. All that was in my regional language "Gujarati". Only other source of information was Television. Discovery channel was just become available during that time, and I loved to watch programs related to cosmos, animals and technology on it, though I had several fields of interest. That was the time when I came to know about actual contributions of great scientists like Isaac Newton, Albert Einstein, Galileo, Copernic's, Yohannes Kepler and others. I was very impressed by the fame of Albert Einstein in those days because he was represented as a genius by my physics teacher and other media. According to Teachers and Newspapers He was the only person who could use 8% of his brain which was greatest ever in human history.

What I learned about physics was all classical physics or Newtonian physics. Relativity and quantum physics were not given in the physics syllabus in detail. I had no idea about relativity in 11, 12 science. Teachers didn't explain anything conceptual about Relativity. What relativity meant to me was "$E = mc^2$". We were taught about time dilatation and Lorentz transformation but just as a short topic and we didn't teach logic behind it and its origin. Relativity remained a mystery for many years for me. I read the first book on relativity for the first time in 2023 at the age of 42. During school days I heard one quote from my friend "Remember M, Remember E, put them together and remember ME". Though it was not written in that meaning by writer and we were not aware of the writer, I found some relation of M, E and ME with physics and fundamentals. Collecting information from multiple sources and keeping them in mind and no early discard was my policy. I always tried to relate Philosophy, Spirituality, and Science but without imagination that don't follow logic and cross boundaries of reality. At the end of the higher secondary education, I was aware about the vastness of universe, size of galaxy, dangerous nature of blackhole, light as a fastest object, basic concept of time dilation, concept of time travel, faster you move in space, slower the time elapses. There was a large and clear model of the universe in my mind and that was a standard model of physics and relativity. I chose medical science as a course of MBBS for graduation for job security. That was a different stream in science and there was no physics in that stream. I liked biology too and I had no problem in learning about the body and medicines. Though I loved fundamental physics, I hated formulas, calculations and other parts of physics taught to me in 11th and 12th science. I still hate it. I was curious about fundamental science but was not a fan of general practical physics with mathematical calculations based on equations. According to me, that was an engineering thing and not a scientist's thing. Actually, I really hated it. When my physics teacher told us that light is an electromagnetic wave which require no medium for its propagation, then what come in my mind was "This is a blunder". I never expected the possibility of any wave without a medium. Actually, the definition of waves requires a medium. That was the first time I expected presence of something in a vacuum.

While studying MBBS course I further gained information about astronomy and fundamental physics in Gujarati from a magazine named "Safari" compiled by Harsal pushkarna. I bought two books from a store at bus station in my town named "Vismaykarak Vigyan (Amazing Science) part 1 and 2" one in green and another in orange hard covers written by Nagendra Vijay and compiled by Harshal Pushkarna. That was relatively cheap, but that much money was precious for me during those days. That magazine and two books contributed very much to my little knowledge about fundamental science. I became a fan of those books and magazines. I wanted to read and understand Relativity in those days, but I neither searched, nor found any source. During MBBS a miraculous source arose, that was internet. We had paid access to it in cybercafe only. I didn't have a PC until 2006 when I was earning money as a government doctor.

After MBBS, during my job as a medical officer I had internet access at home on my own PC. That was another milestone. A social media site was there named Orkut. I was a regular user of Orkut. It was Facebook like application and there were groups categorized as communities. I found a world level platform in that communities. There was a community named "string theory". Brian was the host in that community, and rarely there was any lines posted with the name "Greene". There were other communities also. I discussed many threads in those days.

I posted a comment on some post which I don't know exactly now but the comment roughly meant, Speed of life = speed in a time +speed in a space =c, meant "We are living with speed of light. That thread drew an attention of the Scientist "Brian Greene" the Author of the book "Elegant Universe" a string theorist. He replied on a thread that he used that technic of vectors in a time and space to explain relativity to his students. He also stated that everything is moving at the speed of light. I saved screenshots of that thread, which I could not preserve.

My understanding of science grew during those days. I learned superficially through internet and tv about some theories like string theory, particle physics, M theory, Quantum physics etc. I started thinking differently with my own explanations for the model of the universe. I didn't know about various theories explaining gravity. Gravity was a mystery for me in those days without any question about it. Explanation by theory of relativity and graviton proved un-satisfactory for me which raised questions about the origin of gravity in my mind. I had rejected those theories. The Mystery was approached by my thought process, and it was wandering in my mind. Bernoulli phenomenon was a hint for me. When a gas-like material of medium passes rapidly through space between two objects, those two objects get attracted toward each other. I got my first theory explaining gravity. But it just stayed for a few minutes only and soon it was rejected by my mind. I argued that if there is something in a vacuum space, and it flows between two objects and creates gravity, then a flow must be there in all directions. Without such flow gravity can only be possible in some special direction, with Bernoulli's principle. I discarded this theory. Suddenly there was a strike in my mind, and I found a mechanical answer to this query. I had solved the mystery of gravity in my mind. It was mechanical push by flow of moving things from all direction to all directions. I explained it to my friend and one of my doctor colleagues. I explained it to my father also. My father agreed with me. The responses of other people were cold. For a few days after getting that solution I was excited to find the ultimate answer. I bought Graff papers and drew some diagrams of push gravity mechanism also. I drew a figure of shadow gravity on a graph paper to find a quantitative equation based on geometry. That confirmed that I was right about the gravity and nonblank nature of space. But that joy didn't remain longer when I discussed the explanation of mechanism of gravity on Orkut, a person named Zefir in a community string theory, explained it to me as mechanical push gravity or shadow gravity by Nicholas Fatio De Duillier. That was heart breaking for me. I searched on the internet for that theory, that was same as thought by me. Even Diagram of push gravity drawn by me, was nearly same to the diagrams by Nicolas Fatio but with less details. That followed a long pause in my mind. I found that mechanical push gravity theory was rejected twice by mainstream science. But still I was confident that it could be the only possible answer to solve the mystery of gravity.

The word Matter was dissected by my mind for meaning. Its literal pronunciation in Gujarati was "mâ te ra" could help in getting a meaning "that's why it remains sustained and stable". This word provided me confidence in believing that, Matter is not self-stable, it is made clustered and getting its stability as a cluster by some mechanism. That mechanism may be the push from the surrounding and that surrounding cannot be an absolute blank, but it must be something which could not express itself as a material thing. I came to know also that mechanical push gravity theory was rejected by mainstream science. Gradually that theory had its evolution in my resting mind, moving particles were replaced by tides in a medium. There was a name for a medium in physics. A luminiferous Aether. I became a believer of Aether. But my Aether was not just luminiferous but also a liquid-like material with flows of energy to produce gravity also. I was not sure about its state as a liquid like or a solid like or something else. But I was thinking of a medium that could carry waves. When first time I heard in the Eleven-twelfth science about light as a self-propagating wave without medium and wave particle duality of the photons, I rejected those concepts at a time. I argued that there must be a medium for the propagation of any kind of waves. But that era was of a poor knowledge sources for a normal student in India. After getting on the internet, I started searching for those topics again, discussed those topics in Orkut. I found most thinkers were not in support of relativistic view but were in support of Aether, a hidden unexpressed medium. I wrote my first theory on a model of the universe as an essay "Journey to the Aether". That was around 2008-2009. But I have never published it, that was like an essay of 3 to 4 pages in word document without any calculations and with only some arguments.

I didn't know anything about research papers or book publications at that time. I did not even intend to publish it. But I was overconfident that I was the only one who knew everything properly about fundamentals (☺). I thought that I was the topmost critical thinker because I was not aware of the real hurdles in making an acceptable theory. I thought believing in a concept is all and there is no matter if you can prove it or not. Then there was a dark era of active interest in fundamental science in my life, when I enjoyed my life, got married and enjoyed life by staying outside my mind. I got few transfers in my job, meanwhile a giant named Facebook engulfed my loved one social media Orkut. That was disappointing for me, I tried to recover that app and tried to get it working again on my smart phone. But all in a vein. Groups in Facebook couldn't become an option of communities of Orkut in those days. Internet speed grew from dialup connection to broadband. I had a PC, a laptop and smartphones. But Orkut was not there. I missed it a lot. You-tube arose like a great replacement for tv channels. I used it to know reputed theories. But during my job in the tribble area, I didn't find proper environment for thinking in the science field. I had to lose one of my hobbies in a harsh job. It was 2016 to 2020, I was a married guy without a kid, having facility of four-wheeler and a company of a good wife and good free time. I enjoyed journeys and visited many nearby places, hills, temples, waterfalls, jungles and beautiful places. Those were cool days with some problems and obstacles also. The environment at my job place was not good, I had many serious troubles because of my strong and strict nature and temperaments. My family suffered a lot. I lost my father in 2018. A corona was there in 2020, We did not get any break in a job or an option like work from home, but due to lockdown our workload decreased to some extent resulted in a relief for me. My wife conceived in that peace-full interval, and we had a cute handsome male child in 2021. I was preparing for entrance exam for further studies, but family responsibilities were my priority, and I left those tries in-between. In 2022 I started writing my theories again. Meanwhile I got transferred near to my hometown. I started writing. I wanted to write on Aether model, I stuck at the point that, Relativity is against the existence of any kind of Aether. I downloaded a book on relativity by albert einstein from the free website. Started reading it. It was an English translation of his book in German. It's language and description were miserable and not easy to understand for a person like me who was not fluent in English. I took help from google and translated that book in a Gujarati with many efforts over two months. I thought I understood it properly. I watched many videos related to relativity on YouTube and also read many threads on the topic on various websites. I could grasp the concept, but I found it illogical. I also found a hidden acceptance of Aether in general relativity as a reason behind spacetime matrix. There was a strong argument that, if there is nothing in space then what is geodesic and a wrapped space. The relativity seemed weird to me. I inquired about its foundation and found many breaks in logic. I published some research paper in a journal IJSR and also published a book "Relativity Refuted." Thus, I prepared well to make a ground for my Aether theory. There was a misunderstanding in understanding about time dilation in relativity, and I described only relative time dilation with two observers in my book. I published an article in a science journal about mistakes in the use of Lorentz transformation and limitation of Lorentz transformation after publishing my book. Because Aether ruled out in special relativity, was a substance between masses and not a cause of mass itself, rejection by special relativity had no relation with Aether or Brahm I wanted to describe. I wrote most part of this book before self-publishing a book on Relativity in March 2024. My goal was to publish both books simultaneously, but I had time crises and busy schedule in my government job, spending minimum 12 hours in a day continuously on duty and much more. Actually, we have to do 24 hours shift duty with a whole day on duty and a day off after 24 hours of duty. During such a busy schedule my wife had a second pregnancy and some health issues. Because of her poor health, I temporary lost my only support, and that resulted in the delay in preparing this book for publication. Experience of self-publishing first book on relativity was not quite good because I had hard time in collecting ISBN number and was not able to make available a book for sale on leading website due to unavailability of the GSTIN number. Misunderstanding about time dilation ruined the book. Still there was a good description of many topics. April to July 2024, four months remain disappointing for me. I haven't done any notable work on the book during these four months. A few months after the birth of my baby girl, I got back to writing and continued preparing this book in my free time at night duty while on job, but over-burden work allowed only a few sessions in a month. Later There were some quarries about electricity, light, inertia and few other things which were not solved in my mind properly. So, there was a pause. I transferred it to my unconscious mind for a solution with a time.

What are some logic and observations that led me to this theory?

I found that something was missing there in fundamental science.

1. I rejected the hypotheses very early that a light or EM waves could be a self-propagating wave without a medium.

2. Redshift: redshift of a light coming from far galaxies is a sign of tiredness or transformed light which is an indication toward existence of medium for light in a vacuum. The effect of gravitational force on a spinning heavy metal wheel allowed to settle down on a flat surface, showing fastening of movement by conversion of potential energy into kinetic energy drew my attention at a very early age in childhood while playing with heavy metallic wheel used by my father as a tool for his watch repairing-kit. It is known as "Watch Balance holding wheel". I found similarity in light wave transformation in Brahm and gravitational transformation of potential energy of wheel in kinetic motion. My

argument against expanding universe theory becomes stronger with this observation and becomes confident in my way of theory.
3. When I thought about the existence of a solid with abrupt margins inside an absolute nothing, I literally hesitated to accept this as a pure fact. The general behavior of a cluster of particles with energy is to disperse, but matter is not dispersing. Understanding of gravity as a push provided an answer for that, but it was pointing toward existence of motions and dynamics in a vacuum around any stable matter. I could not accept graviton because theoretically it works against newtons third law of motion which laws are the base for the common sense. I could not accept strings as primary structures because such vibrating strings must interact with surrounding strings and that can ruin the specificity of any string and that made this possibility impossible. Two vibrating solid filaments without any field stay nearer to each other and don't get smashed with each other, that was impossible in my imagination. Queries about stability and origin of strings didn't allow their acceptance as a fundamental units. Relativity was never an option because it doesn't say anything about the vacuum or gravity. Curved spacetime was not the answer. How curvature is formed could be the answer but that was not explained anywhere in relativity. Mathematical errors in calculation of time dilation and length contraction were not acceptable at all. So, I didn't find any plausible explanation of gravity other than mechanical push gravity. This concept led me to Brahm theory to explain push gravity. It was hard time imagining a matter inside Aether. I am not a physics graduate, and I was not aware of various Aether based theories in detail. Vortex as a primary mass forming torus unit was a gradual outcome when I imagined a ring like vortex formed by two or more drop like waves of Aether around a common centre in a vibrant Aether with pushing energy flow. It resulted in the formation of a galaxy like torus structure with a central jet, spinning massive disc and halo forming a field around it. That was a perfect answer for the something within Aether which was accepted as a base of mass. Formation of rest of theory is just a logical extensions with multiple thought experiments. Some modern findings like a quantum vacuum also helped to support my belief. Though I found a relation of topics of Brahm theory with Indian mythological characters and paintings, I find these relations only after understanding those topics by myself. So, I can't consider it as a source of pure knowledge, but it helped in gaining confidence on my way and also helped in developing some ideas. As far as I

wrote the theory, I found related images and concepts in Indian mythology with self-imagined meanings. Electricity was one of the hardest part to be understood properly. I know that still my mind is working on this theory, I decided to publish it at this stage of immaturity of the theory because I don't want to keep it with myself after such progress in the theory. I want to share it with other thinkers also. I don't think about rejection of the theory, because I know that this could be the only conceptual answer to curiosity about fundamentals.

17. Glimpse of General belief in a science including standard model of physics and other theories.

Cosmos is the word used to describe the world at large scale where stars, planets, galaxies, nebulas, interstellar space and other large structures are subjects of concern. Nuclear physics is concerned with fundamental forces and fundamental particles.

There are many theories that may represent cosmos and physics in other way, but here I am going to gather common beliefs accepted as important beliefs or proven things in a mainstream science.

Matter:

Science knows that there could be many states of matters. But matter is made up of atoms mostly. Those atoms have subatomic particles forming nucleus and rest periphery of the atom. Electrons are things orbiting nucleus and they follow some rules of quantum mechanics. Electrons are there but not like a dots or globe of mass orbiting nucleus, it is a cloud like existence in which there are some possibilities to find electrons at some place around nucleus, which's probability could be calculated with Schrodinger's wave function equation. So, electrons are elementary particles and may have a superpositions. According to Heisenberg's uncertainty principle, one can either measure exact position or momentum of quantum particles, measuring one decreases accuracy of measurement of other property. Nucleus is made up of two major particles, protons and neutrons. These two are not same. Protons show attraction toward the negative terminal in the electric field, so, it is considered as a positively charged particle while neutron is considered as electrically neutral on the basis of its neutral behaviour in electric field. Electric and magnetic fields are made up of unknow things, which are curved geodesics selective for charged particles and not for other things. Modern science theorized that these protons and neutrons are not the fundamental end particles, but they are further made up of quarks. Few types of quarks like up quarks and down quarks are theoretical imaginations to fit in mathematical calculations of some observations. Each of proton and neutron is made up of 3 quarks. Charge of proton is positive because it is made up of two up quarks and one down quarks. Charge of neutron is 0 because it has two down and one up quarks. Few recent studies claim about the contribution of 5 quarks in a formation of proton instead of classical 3 quarks. Electrons are considered as fundamental particles with high stability and half-life more than billions of years. Recent experiment on Graffin in quantum physics showed half electrons also, a half of fundamental particle!

Fundamental forces:

Quarks are bound to each other by a hypothetical theoretical mathematically predicted particle gluon, which mediates attraction between quarks. These interaction forms a force what we know as nuclear strong force. Electrons and quarks are known as elementary particles and categorized as fermions. Weak force is the force that mediates radioactivity. Its range is very short, and it is limited to subatomic distance less than the size of proton. Electroweak theory explains the weak force. Unlike strong force weak interaction is mediated by exchange of three types of bosons W+, W- and Z bosons which are also mathematical predictions. There is a no mechanical explanation of this interaction. Quarks are described as composite particle with six different flavours. They can exchange flavours with one another. It is theorized that such exchange explains the radioactive process! I am not going to elaborate on each of everything here.

There are two other fundamental forces. Electromagnetic force and Gravitational force. Electromagnetic force is mediated by electric and magnetic field and interaction of charged particles within this field. Basic concept is that there is an attraction between opposite charge and repulsion between same charges. There is nothing like a mechanical explanation of these concepts and there is no concept about what actually a charge is. Actually, science theorized too much without proper realistic concepts. When there is an imagination called a field, there is a no further explanation of fundamentals units creating these fields.

Gravity

In a modern Einsteinian Relativistic science, gravity is accepted as an effect mediated by the ability of mass to bend spacetime around it and forming geodesics or curved spacetime matrix. It is described as an effect. A mass can create a curvature in a no-Aether spacetime-matrix surrounding it, that bends that blank space, which guides another mass by forming geodesics. How mass creates these curvatures and in what it creates a curvature are the unanswered questions till date. Einstein tried in vain to reintroduce Aether as a neo-ether which was rejected by science community due to its incompatibility with the special theory of relativity. Actually, theory of general relativity demands the existence of Aether and Brahm.

There are some points on which modern physics have not taken any stand but left it unanswered for any possibilities, as described below.

1. An empty-looking space or vacuum is actually nothing or there is something there which doesn't express itself in a detectable manner. But science accepts that there is an existence of various fields in this vacuum like a higg's field, gravity forming field, etc. What carries these fields and how it has such properties is not theorized at all. The Kasimir effect, long-distance force acting through space, enormous life span of stars etc. demands the existence of something in a vacuum but modern science is still not accepting the existence of Aether. Space is capable of producing electric and magnetic fields within itself as the light passes through it. This is an excepted concept about medium-less self-propagating waves having property of wave particle duality.

2. Light: If the photon is a wave or light is a particle, it is not clear. Still, it is also not clear if it has a mass or not. It is considered as massless particles, but math shows error when we try to use 0 mass in classical physics. Relativity tried to explain it, but it is not a theory with proper logic according to me. Still the possibility of light as a mass is not ruled out.

3. Science is changing its view on the model of atoms continuously, there is no description about the real difference between something and nothing. The word "fundamental particle" reveals that science believes in absolute solids. Still science says nothing about the existence of non-spinning most fundamental particles. Science doesn't know if the quarks really make nuclear subatomic particles or not. Science doesn't

know if the proton and neutrons are spinning or not. Science doesn't understand the phenomena of spontaneous formation of elementary particles from nothing and spontaneous disappearance of them in nothing. Actually, there is no method to count the number of electrons spinning around nucleus, the number shown are just assumption based on electric charges and atomic mass of the atoms. A science can detect many things about the atom, these things are "it is a mass which has a more solid centre and less solid near empty surrounding field. This mass has peculiar interaction with other masses; it has many properties like a colour and absorption spectrum. It interacts with light. It has energy. It can carry momentum. It has inertia. It has different types based on its mass. They can form different types of elements, they could be in a charged and neutral state, it may be in a calm or exited state, it may transmit electricity or charge, it could be fused with other atoms, or it could be broken by bombarding neutrons on it. etc."

So, in reality science doesn't know what the atom actually is. When science talks confidently about the exact nature of an atom, it is actually an overemphasis only.

View of Modern science about large scale of cosmos:

It is believed that there is a blackhole in the centre of each spiral galaxy. But it is just an assumption. It is not confirmed. Deficient mass in the universe points toward the existence of dark matter and dark energy, but all that are un-explored sectors of modern science. Even the theory of the big bang is not confirmed with evidence. Expanding universe theory also has many controversies and it is actually a premature conclusion based on one or two characters in observations and ignoring other possible possibilities. Findings of the JWST stand against the assumptions of big bang theory. It found mature galaxies at a large distance which don't fit in a big bang model. It suggests different origins of the galaxy due to its maturity. In simple words, that galaxy ages more than expected. So Big bang model becomes controversial day by day.

Star energy:

Fusion reaction is assumed as a source of all the energy and heat production in the stars, this is a pure assumption and not a result of calculations. Heat at the core of the earth even after billions of years of time passed after the formation of the earth is difficult to explain by modern science. Science doesn't know if this is in correlation with Insulating capacity of earth material with preserved heat inside the core or not. Science believes that the core of the earth is losing heat at a very slow rate, but there is no evidence that if it is still losing heat of the core with decrease in the core heat or not. Scientists also believe that spinning core is responsible for the magnetic field generation, and it means that if core stops spinning due to further heat loss than there will not be a magnetic field at all, and earth may have a fate as planet without life. Science only has assumptions and theories about the formation of elements in the world. Science doesn't know what really happens over a very long period of time at the star system like our solar system. Science doesn't know why big mass clusters like galaxies tend to form a disk like planar structure. Science doesn't know why there are energy fountains at the centre of quasars. Relativistic explanations are against the relativity itself which don't allow any light to escape from blackhole. Science doesn't explain why black holes do not behave like a large vacuum cleaner and why we are not observing any such blackhole larger than galaxies cleaning the space. Science lacks internal co-relations between various accepted theories also. Reality is actually puzzling modern science intensely. Theories are full of paradoxes. Science can't understand the reason behind quanta, pockets of energy or specific sizes of elementary particles.

Accepting final limits of time boundaries and size limits for whole existence itself indicates poor restricted thinking.

18. Relevance with Indian mythology.

Meanings of the Indian mythological characters and paintings in relation to Aether theory according to me.

Torus:

Two-word torus and taurus have a similar pronunciation but different meanings. Torus is a structure in an Aether theory, which has similarity with the God and mythological character "Natraj" or "Shiv" in the Indian mythology. Image of the Natraj represents following things.

Position of hairs, hands and legs represents a spinning state with working centrifugal force. Fountain on head represents central energy flow or even an electric flow. Black colour represents mass or a material thing. The word Shankar has a sub-word nakkar, which represents a solid material. Serpent in neck represents a push gravity and smaller serpent on scalp hair around central energy flow represents less or weak gravity.

Leather clothing represents a mosaic Nature which indicates further smaller units arranged in mosaic forms. Position placed on himself suggest self-stable or floating in a vacuum status. Trishul represents poles. Streams of energy flows represents magnetic field. Damroo represents an hourglass time clock that represents relation of time with the motion of the torus. Taurus known as Nandi is a chosen animal for shiva which represents relation to word torus. In one of the Hindu literatures, it is said that shiva is not different from Nandi, that shows relation of Shiva with taurus. It mentioned in a shiv puraan that Bhagvan shiv blessed Shilad-muni with a boon to get Bhagvan shiv as his own son. Nandi was a son of Shilad-muni who was considered as an Avatar of Shiva himself. Shiva is also known as a

God and Ishwar also. So, this represents the main character in mythology and main structure in a Brahm theory.

Some incident represented in paintings like burning by stream from third eye represents a lesser or gamma burst. Trapping "stream of Ganga" in his hairs or Formation of the "stream of Ganga" from his hairs at the centre of the head, from a pole, represents a jet of quasar, or an electric flow also. It also tells that stream is already trapped in Torus and occasional release is equal to just a fraction of amount of trapped stream. Though different meaning can also be extracted like shiva is a blackhole who can trap energy within itself, and quasar jet are similar to stream of Ganga from his hairs. This is also a relevance with blackhole not predicted by this Brahm theory.

Furthermore, meaning in relation to electric circuit and electricity and battery could be extracted from the painting of "Chhinnamasta devi". Kamadev represents relation with cathode rays. Samudra Manthan shows a personification of the dynamo working model. Painting of Shri Vishnu in a Sir-Sagar with lord Brahma from his umbilicus shows an Aether, with universe as a manifestation within itself. There is relation between words also, like atom and Aatma, Aether and Ishwar. Energy is defined with the word Shagti as a motion, Shankar word represents a solid nature. If you have good knowledge of English, Hindi and Gujarati, you may extract so many interrelations and meaning woven within words. Sameer, Rashmi, Dhananjay these words represent aethereal waves or particles. Shesh serpent represents gravity. Image of Ardh-Nareshvar represents a mixed existence formed with mass-energy dual existence. That shows that mass is nothing but energy only. That also shows cyclic nature of formation and annihilation of the material world in a random energy form. The last cosmic dance before the end of the world Pralay is described as a Tandav, dissected by me a The-End- ow.

Why am I representing such point of view here in this book? Any of these could be a co-incident, but all of them together in such a strong relations can't be a co-incident When the Advait vâd is described in Hindu texts as a philosophy that has close relation with the Brahm theory represented in this book. This strongly suggests an existence of Brahm theory and knowledge of Brahm theory in ancient civilization. What could be the source of such knowledge is a mystery for us.

There are many possibilities, each of them indicates poor reach of our present knowledge about existence of intellect in ancient time.

19. Criticism of Relativity and Relativity in relation to Brahm.

Generally, Relativity is considered as the Special theory of Relativity and General Relativity proposed by the physicist Professor Albert Einstein. These theories show relation of relative times and lengths in relation to relative motions of objects with one another. It also defines Gravity as space time curvature. In the Special Theory of Relativity, relative motions of two objects in relation to each other plays crucial role and that decides relative time, relative mass etc. There is a no preferred frame of reference or universal frame of reference in a special relativity with rejection of any kind of Aether. But still acceleration is used to choose inertial moving reference frame and that decides whose time would dilate in relation to whom. It could also be determined by top view of relative motions. For example, if we spin around then we would find that the world is moving. In that case relative to spinning observer, far objects move with the speed greater that the speed of light c. This indicates that it is the observer who is spinning and not the world is orbiting around him because relative speed more than c is not possible according to the special theory of relativity. Thus, there would not be a time dilation for apparent motion of the apparently orbiting objects.

Lorentz transformation used interchangeability of subjects as permitted by one of the two postulates in special theory of relativity. But it is not allowed in practice. If the perspectives of subjective experience of both subjects are included in the situation by considering both as moving in relation to one another by using subject interchangeability, then it would create a paradox. When that subject interchangeability is applied without restrictions that creates a Twin paradox and predicts non-realistic results. Albert Einstein explained that paradox by showing a way to choose moving reference frame between two relatively moving things. He suggested acceleration as a key to decide a moving reference frame in which time dilates. But still this explanation doesn't explain the situation where both reference frames feel acceleration. In that case where both subjects are moving toward each other and both felt same acceleration, Lorentz transformation can't remain applicable. If we imagine one object at a point at a centre of a path between those two moving objects, then both moving objects should show time dilation in relation to that object and no time dilation in relation to each other. This shows that there is a no time dilation when two objects move equally in relation to each other. In a Lorentz transformation, the mathematics used to show Time dilation is improper that seems illogical. Lorentz transformation is described ahead in this chapter, and this point is covered there.

In General relativity a property to form a curvature in a space-time is imagined as a power of mass itself without any dependency on surrounding space-time. This provides a limitless vicious nature to gravity that enables mass present in the enough quantity to collapse itself with its own power and forms blackhole with singularity. Origine of the power of the mass to form gravitational field is imagined within mass itself, this results in such vicious gravity because now mass is forming gravity and there is a no restrictive role of surrounding. These two theories, Special and General theories of Relativity proposed by Professor Albert Einstein is modified in relation to Brahm-Aether described ahead in this chapter, that also provides a time dilation. Time is described as a fourth dimension but in relation to Aether and not in relation to one another for moving objects. This suggests that accepting Aether or Brahm is not contradictory to the predictions of the theory of relativity by Albert Einstein but in fact Aether-Brahm could be the only way to explain mechanism and cause behind the time dilation.

Preferred frame of reference in the special theory of relativity:

It is believed that there is a no preferred frame of reference for the relative motions of two objects in special relativity theory. Really is it so? Are two objects free to move with any speed in relation to each other? If three objects are there, are they free to move with any speed in relation to one another within a range of speed from 0 to the speed of light c? Answer is no. They can't move with speed beyond some limit in relation to one another according to the special theory of relativity. Though there is a no preferred reference frame but still there is a range that restricts motions of all objects. Let's see a thought experiment for these.

Figure: 19.1

Suppose there are four objects as well as observers named A, O, B and P. We would like to consider O as a steady observer (not feeling acceleration during this experiment). Now speed of the A in relation to O can't be more than the speed of light c. Suppose A is moving with the speed c in a direction of north relative to O. B is also moving with the speed c in opposite side in a direction of south in relation to O. For the observer A, O would be stand still in the time and it won't perceive any change in O, what could be the observation of B from the observer A? That is an impossible physical system according to special theory of relativity. Let's clear it further. Here B is moving with the speed c in relation to O. Now, consider P as an observer moving with the speed of c in relation to B in a same direction of the motion toward which B is moving. As per the classical physics speed of P relative to O would be 2c, but in the context of the special theory of the relativity relative, speed more than c can't be possible. So, adding P with a speed c relative to B would force a situation to change in a situation where relative speed of B in relation to O must not be c or it should be less than c.

This concludes that we can't add series of objects in a line with speed c relative to the nearby object. This would violate one of the postulates of the special theory of the relativity.

Note: Anything which moves with a relative speed equal to c must lose its mass nature and become energy or radiation wave form. This phenomenon is not taken in a consideration during this thought experiment shown above to understand the point of the thought experiment.

We can say that theory of special relativity concludes that there could not be the existence of mass or anything which could travel with the speed more than that of the light in relation to observer and there could not be addition of any object 2 moving away from the observer also moving with any speed more than zero in relation to the object 1 which is moving with the speed of light relative to the observer. Here all motions are to be considered in a single linear direction. Thus, a physical system contains something with restricted motions in relation to one another. It seems like that there is the environment adhered to the physical system that restricts the relative motions beyond some limit and thus also restricts the speed of the absolute motions. That environment doesn't allow limitless speed of motions. So, between two-sided possible speed equal to c, there is a reference frame in between those fastest inertial reference frame which can be considered as the universal reference frame.

Observer specific constancy of the speed of light:

There was a complex observation in a Michaelson Morley experiment which was intended to find a drag by aether on the light propagation.

The observer specific constancy of the speed of light is not easy to understand. Actually, it is so because it is a poorly defined concept in itself.

Look at the figure 19.2below.

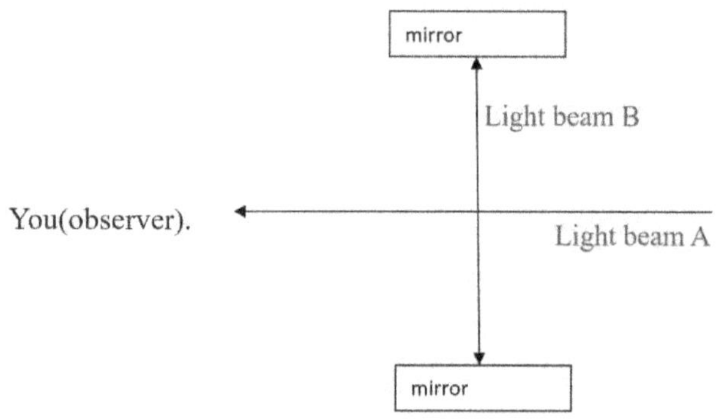

Figure: 19.2

There are two beam of lights A and B. The whole assembly is put in a ship moving with a velocity v in relation to you in horizontal direction.

According to the special theory of relativity there is a relative time dilation for that ship due to its relative motion in relation to you. There would be some different time environment in the ship relative to you. Now both beams of lights are also moving with the ship in relation to ship and also in relation to you. Let me discus their motion in relation to you only. A is travelling in a direction toward you with the speed of light. Its speed relative to you must be c. Now B is also traveling with the speed c in relation to you according to the Special theory of Relativity. The path of the beam B would be of zig-zag pattern due to bouncing of light beam between those two mirrors shown in the figure. After a passage of time when ship travels some distance then the beam of light B will remain at same place bouncing between two mirrors near to beam A. Here if the speed of light of beam B would be equal to A for you than it must be left behind during this travel. But that doesn't happen. Relativity may explain this by saying that there is a time dilation when you see bouncing beam B and there is no time dilation when you see beam A. Due to direction of motion this thing occurs. But we can see here clearly that light beam B is travelling faster in relation to light beam A to stay in the assembly between those two mirrors along with light beam A. This can be compared to the arrangement with interferometer in a Michaelson Morley's experiment where in relation to the space two beam of light travels in two different directions where beam traveling in a zig-zag path travels more distance in relation to space compared to beam traveling in a straight line. After some time both light beams in the interferometer meet at a point which is at equal distance for paths of lights in relation to interferometer from the source of light in a Michelson Morley's experiment and forms the interference. So, where is observer specific constancy of the speed of light here? It is evident that speed of the assembly is being added to the vertical beam, and it get dragged with the assembly. This is explained as two components of light propagation in the "chapter 8 light" in this part of the book. There is a no need to include time dilation to understand the result of the Michaelson Morley's experiment. Obviously, there is a difference in the speed of light for both beams of light. This raises questions about the observer-specific constancy of the speed of light in a vacuum.

So, the conclusion is that there can be an addition to the speed of light if it is hit enough to deviate it from its path. Its speed would remain constant when it is not getting momentum that deviates it from its straight path, when it gets an additional speed in the direction of the deviation or drag. The total speed can be calculated from the vector addition operation. Addition of the speed could be in a perpendicular direction.

Problems with the Special theory of Relativity by Albert Einstein.

Let's see the problems with the Special theory of Relativity, then you will find a theory in relation to Brahm and Aether that explains time dilation.

Postulate 1 of the Special theory of Relativity demands the subject interchangeability for the two-observer moving in relation to each other for the effect of time dilation and length contraction. Let's understand this thing and problems created with this demand of subject interchangeability.

Interchangeability of subjects in a special theory of relativity:

In special theory of relativity by albert Einstein, there was no universal reference frame accepted. There is a concept of the preferred moving inertial reference frame which was introduced later to deal with the twin paradox.

According to the first postulate of the Special theory of Relativity all inertial reference frame should have same and simple law of physics within the inertial reference frame. when two observers A and B move in relation to each other, For A, itself is steady and B is moving, similarly for B, itself is steady and A is moving. If we apply Lorentz transformation, there will be slowing down of the time of B for A and similarly there should be slowing down of the time of A for B.

This subject interchangeability causes paradoxes, and it suggests non-realistic impossible outcomes.

I would like to explain it with example of uncle -nephew relationship.

U is uncle of N

N is nephew of U.

What subject interchangeability states is that application of this relation is same for each of the two subjects in relation to each other.

Means it states that,

For U, U is an uncle of N and N is nephew of U.

And For N, N is an uncle of U and U is a nephew of N.

Means both feels own self as uncle of the other one.

Now is it clear that such relation is impossible, which can be seen in the twin paradox.

Let me explain you twin paradox with the help of an example of two clocks.

S and M are two clocks.

Both are moving in relation to reach other.

For S, S itself is steady, and M is moving so, time in M will be slow down.

For M, M itself is steady, and S is moving so time in S will be slow down.

So according to the Special theory of Relativity with subject interchangeability, M shows less time than S and S also shows less time than M. We can understand that this is not possible. This is a twin paradox.

Suppose S and M are two twin brothers.

When S is on the earth and M is going on a space trip with tremendous speed, and came back to the earth, there would be a relative time dilation for each subject.

S will find that he is aged to an adult and M is still a child because for S the relatively moving subject is M which should show a slowing down of time.

Similarly, M will find that he is aged to an adult and S is still a child because for M the relatively moving subject is S which shows a slowing down of time.

For S there should be a slow ageing of M and for M there should be a slow ageing of S.

When they both meets, there is an impossible controversial situation were finding of both subjects would be personal and impossible. It is not possible that two persons can be elder than each other.

This was considered as the error in the Special theory of Relativity by few peoples, but Einstein tried to explain the situation and favours a preferred moving inertial reference frame based on situation and presence of the acceleration experienced by subjects prior or during the journey. He stated that which one feels acceleration is moving and which one don't feel acceleration is steady and time dilates for moving in relation to steady. So, here that who goes to the space trip will show a time dilation in relation to steady subject. So, in above example S will age to an adult but M won't be due to time dilation for moving subject only.

We know that acceleration is felt for a time when object achieves speed, after achieving speed if it moves with constant velocity then it feels no more acceleration. What factor affects during that constant velocity motion which causes time dilation even after acceleration is no longer remained? This suggests existence of something that decides time dilation in relation to it. That something is not mentioned in the Special theory of Relativity.

Also, there is another way to decide Preferred reference frame which is a type of relative motions. Suppose if an asteroid orbits a planet, then both planet and asteroid are moving in relation to each other but by considering an apparent relative motions of the distant stars and analysing a motion in relation to each other it would become evident that asteroid is orbiting a planet and not a planet is showing an inconsistent variable motion in relation to asteroid. So, with all this application special theory of the relativity states that there could be a time dilation in a moving reference frame and not in a non-moving one. When both are feeling acceleration then there would be a complex relative relation which would decide time dilation more or less in relation to each other depending on their movements considered with effect of acceleration.

All this suggests that in the concept of the Special theory of Relativity, the interchangeability of subject is there and passage of time is defines on the bases of relative motions only but in a practice the special theory of the relativity has to determine one inertial reference frame as steady and other as moving. Time dilation relation cannot be applicable bilaterally. If acceleration affects time dilation in addition to relative rectilinear velocity, it means that there are factors affecting relative relations of two relatively moving objects which are neither included in postulates nor incorporated in the math used in the theory. So, rejection of Aether by special relativity invites explanation of the mechanism by which acceleration affects time dilation.

Mathematical errors in SR:

There are mistakes while deriving time dilation equation and getting co-ordinate transformation with Lorentz's transformation as described below.

The Special theory of Relativity has the two basic postulates which has potential to derive other predictions by using math with those postulates.

Let's see those two basic postulates of special theory of relativity.

1st: - Laws of physics are same and can be stated in their simplest form in all inertial frames of reference.

2nd: - Speed of light is constant for the observer irrespective of the reference frames and relative speed of the source of light.

Lorentz transformation is a method for the co-ordinate transformation in the Special theory of Relativity that replaces the Galilean transformations of classical physics.

I want to point out the mistake in this Lorentz transformation method.

Let's see calculations of Lorentz transformation equation with simple method.

Below, there is a figure 19.2, shown to explain Lorentz transformation.

Figure: 19.2

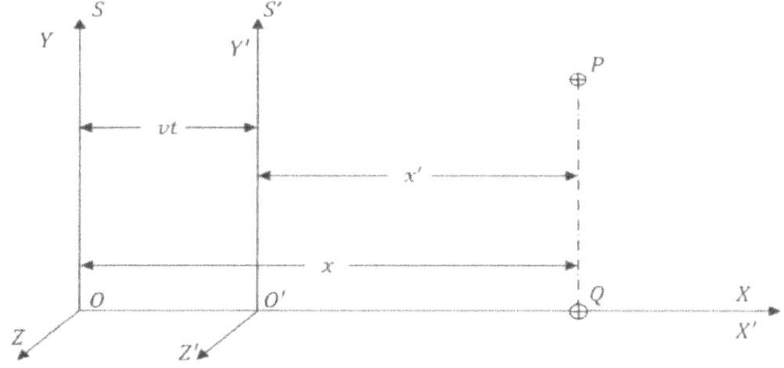

Here two inertial reference frames S and S' are aligned with cartesian planes, which defines co-ordinates of the point $P(x,y,z,t)$ and $P'(x',y',z',t')$ in S and S' inertial reference frames respectively. Here S' inertial reference frame is moving with uniform rectilinear velocity "v" along the direction of positive X axis of S frame. There is no movement of S' along Y or Z axis of the frame S. Both frames are aligned in a way that O and O' co-insides at time $t = t' = 0$. Here P is the point selected randomly in the stationary reference frame S toward Positive X axis side of the reference frame. Q is the reference point on X axis in the reference frame S with co-ordinates $(x,0,0,t)$. In a time "t" photon travels perpendicular to Y axis from the point P for a distance equal to $OQ = x = ct$ and meanwhile during the same time t, S' frame moves a distance vt along positive X axis.

Note:

(One thing to-be noted here is that, though we are transforming co-ordinates for P, we are doing all calculations for Q. Understand this, for the conveniency of the calculations we have assumed that $y' = y$, and $z' = z$. Means relative movement of P for S' frame is only on X axis. So, there will not be any change in y and z during transformations, so they will not appear in calculations, only x will appear. We will define a time co-ordinate for P in a way that x will be equal to ct. Here OP is not equal to ct, but OQ is equal to ct and all calculations are for Q and it will be applicable as calculations of x co-ordinate of P while transformations.

We will not use confusing relation as below,

$$x^2 + y^2 + z^2 = d^2 \qquad here\; d = OP = ct$$

$$x^2 + y^2 + z^2 = c^2t^2 \quad \dots\dots\dots\dots\dots\dots\dots\dots\dots\dots(C)$$

Why are we not taking these relations?

Because here we are able to calculate for x only and we will have to take $y = 0, z = 0$ to derive constant Υ.

When we do that, these relations would become,

$$x^2 + (0)y^2 + (0)z^2 = c^2t^2 \, ,$$

$$x^2 = c^2t^2$$

So,

$$x = ct \qquad\qquad (no\ negative\ value\ for\ time.)$$

If we put this value in equation C,

$$c^2t^2 + y^2 + z^2 = c^2t^2$$

It will yield $y = 0$, and $z = 0$, means we are restricted to choose a point on the X axis only for calculations, after getting transformation for x, we can apply it to co-ordinates for P in S' frame, but the equations for calculation of x, must consider $y = 0$, and $z = 0$, otherwise that won't work.

So, Lorentz transformation equations are for Q and not for P. We just apply transformation of co-ordinates of Q to the x co-ordinate value of P. So, there will be Q only in calculations ahead.

Relation between O and Q is same as relation between P and nearest point on Y axis of S frame.

Relation between O' and Q is same as relation between P and nearest point on Y' axis of S' frame.

So, $x = ct$ is applied wherever there was a need in calculations,

We are free to choose a point on the X axis so, there should not be an objection for calculations for the point Q.)
End of the note.

From the figure 19.2

x is the co-ordinate of P on the X axis.

$$x = ct \qquad\qquad ...... (19.1) \qquad\qquad (read\ "note"\ above)$$

And

$$x' = ct' \quad \ldots\ldots (19.2) \quad \text{(as per the postulate 1 of SR)}$$

From the figure 19.2, Newtonian relations can be,

$$x' = x - vt \quad \ldots\ldots (19.3)$$

and

$$x = x' + vt \quad \ldots\ldots (19.4)$$

If we applied relativistic relation in eq. 19.3,

$$x' = Y(x - vt) \ldots\ldots\ldots\ldots\ldots\ldots (19.5)$$

Here Y is the constant.

Now using first postulates of the special theory of relativity.

(using subject interchangeability)

$$x = Y(x' + vt') \ldots\ldots\ldots\ldots\ldots\ldots (19.6)$$

I have an objection for this equation 19.6 and this is the root for all errors in the Special theory of Relativity. Applying such constant is not logical.

For now, continuing the calculation.

Putting the value of x' in the equation no (19.6) from the equation no (19.5)

$$x = Y(x' + vt')$$

$$x = Y(Y(x - vt) + vt')$$

$$x = Y^2(x - vt) + Yvt'$$

$$x = Y^2 x - Y^2 vt + Yvt'$$

$$Yvt' = x - Y^2 x + Y^2 vt$$

$$t' = \frac{x - Y^2 x + Y^2 vt}{Yv}$$

$$t' = \frac{x - Y^2(x - vt)}{Yv} \quad \dots\dots\dots\dots\dots\dots\dots \ (19.7)$$

Now using second postulates of special theory of relativity.

Distance OQ is Equal to x and is equal to ct

Distance $O'Q$ is equal to x' and is equal to ct'
So,

$$x = ct$$
$$x' = ct' \quad \dots\dots \quad \dots\dots \quad \dots\dots \ (19.8)$$

Now putting the value of x' & t' in this equation (19.8)

$$x' = ct'$$

$$Y(x - vt) = c.\frac{x - Y^2 x + Y^2 vt}{Yv}$$

$$Yx - Yvt = c.\frac{x - Y^2 x + Y^2 vt}{Yv}$$

$$Yx * Yv - Yvt * Yv = c(x - Y^2 x + Y^2 vt)$$

$$Y^2 xv - Y^2 v^2 t = cx - Y^2 cx + Y^2 cvt$$

Now putting the value of $x = ct$ in the above equation

$$Y^2 xv - Y^2 v^2 t = cx - Y^2 cx + Y^2 cvt$$

$$Y^2 ctv - Y^2 v^2 t = c * ct - Y^2 c * ct + Y^2 cvt$$

$$Y^2 ctv - Y^2 v^2 t = c^2 t - Y^2 c^2 t + Y^2 cvt$$

Cancelling $Y^2 cvt$ on both sides.

$$-Y^2 v^2 t = c^2 t - Y^2 c^2 t$$

$$c^2 t = -Y^2 v^2 t + Y^2 c^2 t$$

$$c^2 t = Y^2 c^2 t - Y^2 v^2 t$$

$$c^2 t = Y^2 t (c^2 - v^2)$$

$$c^2 = Y^2 (c^2 - v^2)$$

$$Y^2 = \frac{c^2}{c^2 - v^2}$$

Dividing above and below with c^2

$$Y^2 = \frac{1}{1 - {v^2}/{c^2}}$$

$$Y = \frac{1}{\sqrt{1 - {v^2}/{c^2}}}$$

Now putting the value of Y in the equation no (19.7),

$$t' = \frac{x - Y^2 (x - vt)}{Yv} \quad \text{...... ... ...} \quad (19.7)$$

$$= \frac{x - \dfrac{c^2}{c^2 - v^2}(x - vt)}{v \dfrac{c}{\sqrt{c^2 - v^2}}}$$

$$= \frac{\dfrac{c^2 x - xv^2 - c^2 x + c^2 vt}{c^2 - v^2}}{\dfrac{vc}{\sqrt{c^2 - v^2}}}$$

$$= \frac{c^2 x - xv^2 - c^2 x + c^2 vt}{vc\sqrt{c^2 - v^2}}$$

$$= \frac{-xv^2 + c^2 vt}{vc\sqrt{c^2 - v^2}}$$

$$= \frac{c^2 vt - xv^2}{vc\sqrt{c^2 - v^2}}$$

$$= \frac{c^2 v(t - {xv}/{c^2})}{vc\sqrt{c^2 - v^2}}$$

$$= \frac{c(t - {xv}/{c^2})}{\sqrt{c^2 - v^2}}$$

$$= \frac{t - {xv}/{c^2}}{\frac{\sqrt{c^2 - v^2}}{c}}$$

$$= \frac{t - {xv}/{c^2}}{\sqrt{\frac{c^2 - v^2}{c^2}}}$$

$$t' = \frac{t - {xv}/{c^2}}{\sqrt{1 - \frac{v^2}{c^2}}} \qquad \ldots \ldots \ldots \ldots \ldots (19.8).$$

Here $v < c$.

So, Lorentz transformation equations are,

$$x' = \frac{x - vt}{\sqrt{1 - \frac{v^2}{c^2}}} \ldots \ldots \ldots \ldots \ldots \ldots \ldots (19.9)$$

$$y' = y$$

$$z' = z$$

$$t' = \frac{t - {xv}/{c^2}}{\sqrt{1 - \frac{v^2}{c^2}}}$$

Inverse Lorentz transformation equations would be,

$$x = \frac{x' + vt'}{\sqrt{1 - \frac{v^2}{c^2}}} \qquad \ldots \ldots \ldots \ldots \ldots \ldots \ldots \ldots (19.10)$$

$$y = y'$$

$$z = z'$$

$$t = \frac{t' + \frac{xv}{c^2}}{\sqrt{1 - \frac{v^2}{c^2}}}$$

What is the postulate 2 used here? it is the observer specific constancy of the light speed. Here we have considered only one situation where there is a motion of the moving reference frame toward the source of the light. What if that reference frame would move away from the source of the light. There would be a negative velocity, and these would result in following equations for the figure shown below which are same as inverse Lorentz transformations equations.

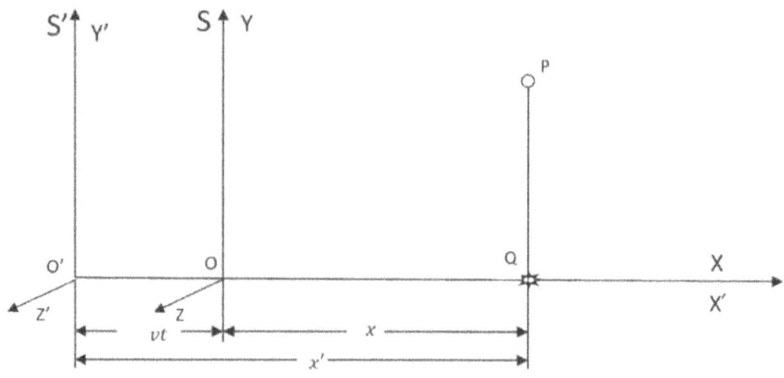

Figure: 19.3

Here only difference is S' is moving away from the P and Q.

Relations would be like,

$$x' = Y(x + vt) \dots\dots\dots\dots\dots\dots (19.11)$$

Here Y is the constant.

Now using first postulates of the special theory of relativity.

(using subject interchangeability)

$$x = Y(x' - vt') \ldots\ldots\ldots\ldots\ldots\ldots (19.12)$$

This would result in,

Lorentz transformation equations,

$$x' = \frac{x + vt}{\sqrt{1 - \dfrac{v^2}{c^2}}}$$

$$y' = y$$

$$z' = z$$

$$t' = \frac{t + {xv}/{c^2}}{\sqrt{1 - \dfrac{v^2}{c^2}}}$$

These are same as inverse Lorentz transverse equations.

Remember my objection for the equation no 19.6

That objection intended to say that it is a root for errors and paradoxes. If we are trying to find relativistic relation by applying observer specific constancy of the speed of light, then adding a constant gamma is wrong logically. There should be a following relation only.

$$x' = x - vt \ldots\ldots\ldots\ldots\ldots\ldots (19.3)$$

and

$$x = x' + vt \ldots\ldots\ldots\ldots\ldots\ldots (19.4)$$

Now,

$$x = ct$$

And

$$x' = ct' = x - vt$$
$$ct' = ct - vt$$

$$t' = \frac{ct - vt}{c}$$

$$t' = t\frac{c - v}{c}$$

Now

$$x' = ct'$$

put the value of x' & t' in this equation

$$x' = ct'$$
$$x' = c.t\frac{c - v}{c}$$
$$x' = t(c - v)$$

Here $v < c$.

So, relativistic transformation equations are,

$$x' = t(c - v) \quad \ldots \ldots \ldots \ldots \ldots \ldots (19.13)$$
$$y' = y$$
$$z' = z$$
$$t' = t\frac{c - v}{c} \quad \ldots \ldots \ldots \ldots \ldots (19.14)$$

If moving reference frame is moving away from the source of the light then,

relativistic transformation equations would be,

$$x' = t(c + v)$$
$$y' = y$$
$$z' = z$$
$$t' = t\frac{c + v}{c}$$

This would show a time contraction.

So, in a case where something moves away first, then came back, as a sum of event there would be following change in a relative time.

$$2t' = t\frac{c + v}{c} + t\frac{c - v}{c}$$
$$2t' = t\frac{c + v + c - v}{c}$$
$$2t' = t\frac{2c}{c}$$
$$2t' = 2t$$

$$t' = t$$

So, when object comes back there would not be any net time dilation or time contraction.

So, really observer specific constancy of the light doesn't yield any absolute time dilation, that is a wrong calculation of the constant Υ that results in such wrong calculation of the time dilation. Mathematics doesn't support it.

Let's see why.

Following equation gives the value of dilated time in moving reference frame,

$$t' = \frac{t - {xv}/{c^2}}{\sqrt{1 - \frac{v^2}{c^2}}} \qquad \ldots\ldots\ldots (19.8)$$

Time dilation:
If we put a clock at the origine of the cartesian plane attached to the moving S' reference frame, then x would be Zero.
Now put $x = 0$ in the equation 19.8.

$$t' = \frac{t - {(0)v}/{c^2}}{\sqrt{1 - \frac{v^2}{c^2}}}$$

$$t' = \frac{t - 0}{\sqrt{1 - \frac{v^2}{c^2}}}$$

$$t' = \frac{t}{\sqrt{1 - \frac{v^2}{c^2}}}$$

$$t' = \Upsilon t$$

So, if the clock moves with a speed equal to v in relation to observer, time t' would be slowed down in a clock as per the equation $t' = \Upsilon t$, which is considered as a relativistic time dilation.

This is calculations according to special relativity.
Criticism:
Criticism About time dilation.
In figure 19.3 if we want to put a clock at the origin of the S' frame.

Then value of the x becomes zero and we get time dilation equation as above.

But we have used in the derivation of the Lorentz transformation that $x = ct$.

So, if we take $x = 0$ then ct becomes zero.

If $ct = 0$ then t must be 0 because c is the constant and it can't be zero.

If t is zero, then t' also becomes zero and no event can happen in a figure 19.3 and all objects remain aligned on one vertical line on Y axis. Time dilation can't be defined in this situation.

So, showing time dilation by taking $x = 0$ is violation of the laws of the mathematics. Even without mathematics if we try to understand that x is a distance that light travels in a time t and if a distance is zero then time taken for the light to travel must be zero. If we don't consider here t equal to zero than we are trying to say that light takes more time to travel a no distance! So, assuming a clock on the point of observer is violating requirements for the Lorentz transformation and it don't yield anything logically sound.

Similarly, calculation of Lorentz Fitzgerald length contraction is also wrong. Within the context of the special theory of relativity also and there should be new calculation if we reject the derivation of the constant Y.

What this means is that time dilation and length contraction are relative thing even within the scope of special relativity and Lorentz transformation and those can be appreciated only when two observers observe object away from them. Superimposing that object on any one of the two observers violates mathematics. If we reject the derivation of the constant Y and accept simple relativistic equations for observer specific constancy of the light, there could be a derivation of relative time contraction and relative length dilation also.

Photon clock showing just a relativistic time dilation is also not justifiable because if we change the arrangement in a photon clock it becomes able to show relative time contraction also. Photon clock type one (used to explain relativistic time dilation) and photon clock type two showing time contraction are explained ahead in this book.

So, Major mistake is accepting following relations true.

$$x' = Y(x - vt) \ldots\ldots\ldots\ldots\ldots\ldots (19.5)$$
$$x = Y(x' + vt') \ldots\ldots\ldots\ldots\ldots\ldots (19.6)$$

This is the mistake at a root. Derivation of factor Y is wrong.

Another mistake is violating limitations of the Lorentz transformation to derive Time dilation and length contraction.

Mistakes in the calculation of the Lorentz Fidge Gerald's length contraction.

Let me show you mistakes in the calculation of the Lorentz Fidge Gerald's length contraction.

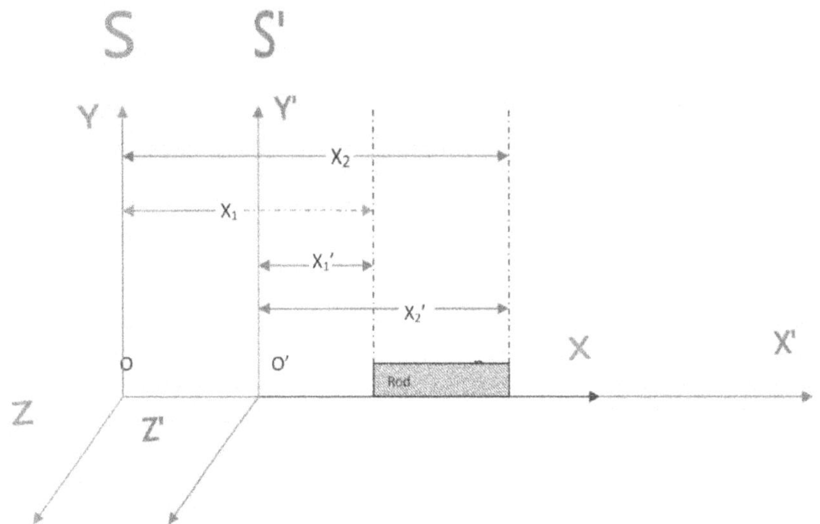

Figure 19.4

S' inertial reference frame is moving with uniform rectilinear velocity "v" along the X axis.

Rod is placed on the positive side of the cartesian plane aligned to the X axis of the S inertial reference frame. In stationary inertial reference frame S, distance of proximal end of Rod from the origin of the reference frame is x_1 and distance of the distal end of the rod is x_2, Proper length of Rod in S is l_0.

In moving inertial reference frame S', distance of proximal end of Rod from the origin of the reference frame S' is x_1' and distance of the distal end of the rod is x_2'. So, contracted length of the Rod in S' is l.

Derivation of the Lorentz Fitz Gerald length contraction using Lorentz transformation equations is done in the special theory of the relativity as follows.

Lorentz transformation equation for x co-ordinate is,

$$x' = \frac{x - vt}{\sqrt{1 - \frac{v^2}{c^2}}} \dots \dots \dots \dots \dots \dots \dots \dots \dots \dots \dots (19.9)$$

$$x = \frac{x' + vt'}{\sqrt{1 - \frac{v^2}{c^2}}} \dots \dots \dots \dots \dots \dots \dots \dots \dots \dots \dots (19.10)$$

In a figure 19.4, a rod is put along the X axis. Length of the rod in relation to stationary reference frame is proper length l_0 Length of rod in relation to moving reference frame is contracted length l

Proper length $\qquad l_0 = x_2 - x_1$ $\qquad\qquad\qquad \dots \dots \dots (19.15)$

Contracted length $\quad l = x_2' - x_1'$ $\qquad\qquad\qquad \dots \dots \dots (19.16)$

Now let put value of (19.10) in the equation no (19.15)

$$l_0 = x_2 - x_1$$

$$l_0 = \frac{x_2' + vt_2'}{\sqrt{1 - \frac{v^2}{c^2}}} - \frac{x_1' + vt_1'}{\sqrt{1 - \frac{v^2}{c^2}}}$$

$$l_0 = \frac{x_2' + vt_2' - x_1' - vt_1'}{\sqrt{1 - \frac{v^2}{c^2}}}$$

But, both length x_2' and x_1' are measured simultaneously at a time t' in moving inertial reference frame S'.

(note: this line is used to derive length contraction traditionally, though this is not possible, but we are continuing for a while to see calculations)

So, $t_1' = t_2' = t'$

So,

$$l_0 = \frac{x_2{}' + vt' - x_1{}' - vt'}{\sqrt{1 - \frac{v^2}{c^2}}}$$

$$l_0 = \frac{x_2{}' - x_1{}'}{\sqrt{1 - \frac{v^2}{c^2}}}$$

$$l_0 = \frac{l}{\sqrt{1 - \frac{v^2}{c^2}}} \qquad \ldots\ldots\ldots\ldots\ldots\ldots\ldots\ldots (19.17)$$

And,

$$l = l_0 \sqrt{1 - \frac{v^2}{c^2}} \qquad \ldots\ldots\ldots\ldots\ldots\ldots\ldots (19.18)$$

Here we can see that for described value of v, l will always be less than l_0

Because value $\sqrt{1 - \frac{v^2}{c^2}}$ will always be less than 1 for $0 < v < c$ value of v.

So, Length contraction is Δl

$$\Delta l = l_0 - l$$

$$\Delta l = l_0 - l_0 \sqrt{1 - \frac{v^2}{c^2}}$$

$$\Delta l = l_0 \left(1 - \sqrt{1 - \frac{v^2}{c^2}}\right)$$

Note: Here I have shown a mistake in the logic applied by putting a note in a bracket, but here the calculation of Lorentz Fitzgerald length contraction is shown as it is described in the special theory of relativity by Professor Albert Einstein. Criticism is ahead.

Lorentz Fitzgerald length contraction recalculated with correction of mistakes in original calculations.

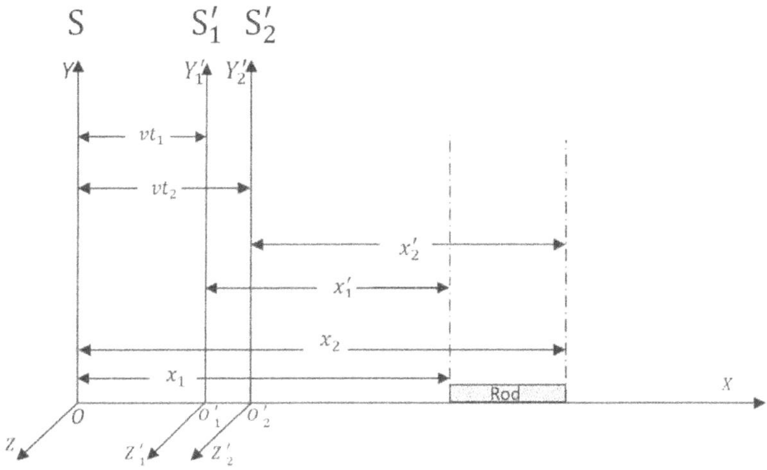

Figure 19.5

Here a cartesian plane is aligned to the steady inertial reference frame S. S' inertial reference frame is moving with uniform rectilinear velocity "v" in the direction of an X axis. Two situations of the S' are shown to get x values for proximal and distal and of the Rod.
Rod is placed on X axis. In a stationary reference frame S, Co-ordinates of the proximal end of the Rod is $(x_1, 0, 0, t_1)$ and co-ordinates of the distal end is $(x_2, 0, 0, t_2)$

Now,
Lorentz transformation equations are,

$$x' = \frac{x - vt}{\sqrt{1 - \dfrac{v^2}{c^2}}} \quad \ldots \ldots \ldots \ldots \ldots \ldots \ldots \ldots (19.9)$$

$$x = \frac{x' + vt'}{\sqrt{1 - \dfrac{v^2}{c^2}}} \quad \ldots \ldots \ldots \ldots \ldots \ldots \ldots \ldots (19.10)$$

In a figure above the Rod is put along the X axis. Length of the rod in relation to stationary reference frame is proper length l_0 Length of the rod in relation to moving reference frame is contracted length l

Proper length $\qquad l_0 = x_2 - x_1 \qquad\qquad \dots\dots\dots (19.15)$

Contracted length $\quad l = x_2' - x_1' \qquad\qquad \dots\dots\dots (19.16)$

Now let's put value of (19.10) in the equation no (19.16).

$$l = x_2' - x_1'$$

$$l = \frac{x_2 - vt_2}{\sqrt{1 - \dfrac{v^2}{c^2}}} - \frac{x_1 - vt_1}{\sqrt{1 - \dfrac{v^2}{c^2}}}$$

$$= \frac{x_2 - vt_2 - x_1 + vt_1}{\sqrt{1 - \dfrac{v^2}{c^2}}}$$

$$l = \frac{(x_2 - x_1) - (vt_2 - vt_1)}{\sqrt{1 - \dfrac{v^2}{c^2}}}$$

$$l = \frac{(x_2 - x_1) - \left(\dfrac{v}{c}ct_2 - \dfrac{v}{c}ct_1\right)}{\sqrt{1 - \dfrac{v^2}{c^2}}}$$

$$l = \frac{(x_2 - x_1) - \dfrac{v}{c}(ct_2 - ct_1)}{\sqrt{1 - \dfrac{v^2}{c^2}}}$$

$$= \frac{(x_2 - x_1) - \dfrac{v}{c}(x_2 - x_1)}{\sqrt{1 - \dfrac{v^2}{c^2}}}$$

$$= \frac{l_0 - \frac{v}{c} l_0}{\sqrt{1 - \frac{v^2}{c^2}}}$$

$$= l_0 \frac{(1 - \frac{v}{c})}{\sqrt{1 - \frac{v^2}{c^2}}}$$

$$= l_0 \frac{\sqrt{1 - \frac{v}{c}}}{\sqrt{1 + \frac{v}{c}}} * \frac{\sqrt{1 - \frac{v}{c}}}{\sqrt{1 - \frac{v}{c}}}$$

$$l = l_0 \frac{\sqrt{1 - \frac{v}{c}}}{\sqrt{1 + \frac{v}{c}}} \qquad here \; v < c$$

$$l = l_0 \sqrt{\frac{1 - \frac{v}{c}}{1 + \frac{v}{c}}} = l_0 \sqrt{\frac{\frac{c - v}{c}}{\frac{c + v}{c}}}$$

$$l = l_0 \sqrt{\frac{c - v}{c + v}}$$

Above equation shows the value of contracted length l.

Length contraction is Δl

$$\Delta l = l_0 - l$$

$$\Delta l = l_0 - l_0 \sqrt{\frac{c - v}{c + v}}$$

$$\Delta l = l_0 \left(1 - \sqrt{\frac{c - v}{c + v}}\right) \qquad (here, \; v < c)$$

In the Special theory of Relativity,

vt_2 and vt_1 are considered same with the argument that x_2' and x_1' are measured simultaneously, so t_2 and t_1 remains same and vt_2 becomes equal to vt_1. So, that gets cancelled during calculations.

But The phrase "x_2' and x_1' are measured simultaneously" is an impossible task for $x_2 \neq x_1$.

Let's understand this.

See the figure 19.2 in the Lorentz transformation.

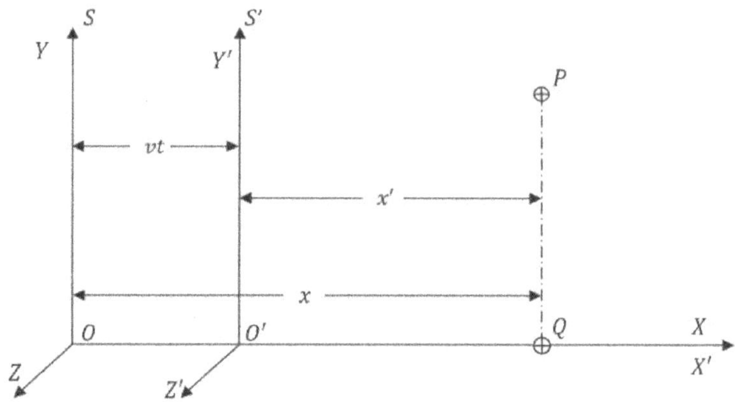

Here, vt is the distance of Y' axis of *the* S' frame from the Y axis of the S frame.
This is the distance that reference frame S' travels along the X axis in a time t.
Now what is this time t?
This is the time taken by photon traveling with the speed c from Q to O and distance equal to the value of co-ordinate x of the point P in reference frame S.
And,

$$ct = x$$

Here c is the constant, so, value of t is depending on the value of x.

As the value of x changes, value of the t will also change.
If we consider the value of v constant,
when we measure x', then vt depends on the value of x.

$$vt = xv/c$$

Means a value of x determines the value of t in ct and that value of t is there in vt which affects the value of vt and thus value of x' also.
So, for,

$$x_1 \neq x_2 \Leftrightarrow t_1 \neq t_2 \Leftrightarrow vt_1 \neq vt_2 \Leftrightarrow x - vt_1 \neq x - vt_2 \Leftrightarrow x'_1 \neq x'_2$$

Lorentz transformation equations are equations which are made to calculate co-ordinates of one point only at one time, where time $t = x/c$. If x varies, then time t will also vary according to the value of x, you can't choose that time t according to your free will for particular x and can't choose same t for two different x values.

So, we can't measure both x_2 and x_1 simultaneously. Because when we apply Lorentz transformation Equation for the point x_2', the value of x_2 will be equal to ct_2 where t_2 will be specific for x_2. In that time t_2 , the moving reference frame S' will move a distance vt_2 , which is also a specific distance depending on the value of t_2 and so, also depending on the value of x_2.

when we apply Lorentz transformation Equation for point x_1', the value of x_1 will be equal to ct_1 where t_1 will be specific for x_1. In that time t_1 the moving reference frame S' will move a distance vt_1. which is also a specific distance depending on the value of t_1 and so, also depending on the value of x_1.

So, the value of vt_2 can't be equal to vt_1, means a position of reference frame S' can't be at same place when we are applying the Lorentz transformation equations for two different points having value of x co-ordinates different from each other.

So, measuring distance for proximal and distal end of the rod simultaneously in a moving reference frame is not possible, because when we apply Lorentz transformation equation value of t' will be different for different x values when velocity v of the moving reference frame will remain constant. So, we can calculate value of x_1' and x_2' for respective value x_1 and x_2 at same velocity v, but we can't calculate them at the same time t.

Let's think about the length contraction described in the book "Relativity, the Special and the General theory by Albert einstein, 1916"

Following is described in that book

-:-

"There, rod is put in a moving reference frame with proximal end on the (0,0). There, time t_1 becomes 0, and time t_2 becomes t and,

contracted length is $\quad l = x_2 - 0 = x_2$

proper length is $l_0 = x_2' - x_1' = Y(x_2 - vt - x_1 + vt)$

And so,

$$l_0 = Y(x_2 - x_1)$$

$$l_0 = Y(x_2 - 0)$$

$$l_0 = Yx_2$$

$$l_0 = Yl$$

$$l = \frac{l_0}{Y}$$

$$l = l_0 \sqrt{1 - \frac{v^2}{c^2}}$$

-:-

So, there we can find contracted length as above. We could not see the mistake in the calculation there because there t_1 is taken as 0 and vt is canceled anyhow by taking $t_2 = t = 0$.

The line is there, "Lorentz transformation value of both end of the rod at time $t = 0$ is as below"

Means in that case, in the book of the theory of relativity, it is mentioned that the distance of the distal end of the rod placed in a moving reference frame can be calculated by Lorentz transformation equation in relation to stationary reference frame S by taking $t = 0$ value in the vt

This is the mistake, if proper distance x is not zero, then ct cannot be zero.

If ct cannot be zero, then vt cannot be zero for v more than 0. So, by using Lorentz transformation equation one cannot take the value of $t = 0$ for the point having value of x coordinate more than zero. So, mistake is also there in the Book of Relativity by albert Einstein.

Please note that I am not agree with the corrected equations shown above for Lorentz Fietz Gerald's length contraction because as described ahead there is a wrong calculation in the derivation of the constant Υ.

What could be the proper equation for relativistic length contraction by accepting both postulates of the theory of the special relativity?

It could be as follow.

relativistic transformation equations are,

$$x' = t(c - v) \quad \dots\dots\dots\dots (19.13)$$
$$y' = y$$
$$z' = z$$
$$t' = t\frac{c - v}{c} \quad \dots\dots\dots\dots (19.14)$$

So, here,

Proper length $\qquad l_0 = x_2 - x_1 \qquad\qquad \dots\dots\dots (19.15)$

Contracted length $\qquad l = x_2' - x_1' \qquad\qquad \dots\dots\dots (19.16)$

Now let's put value of (19.13) in the equation no (19.16).

$$l = x_2' - x_1'$$
$$l = t_2(c - v) - t_1(c - v)$$
$$= t_2 c - t_2 v - t_1 c + t_1 v$$
$$l = ct_2 - ct_1 - (vt_2 - vt_1)$$
$$l = (x_2 - x_1) - (vt_2 - vt_1)$$

$$l = l_0 - (vt_2 - vt_1)$$

Above equation shows the value of contracted length l.

Length contraction is Δl

$$\Delta l = l_0 - l$$
$$\Delta l = l_0 - (l_0 - (vt_2 - vt_1))$$
$$\Delta l = v(t_2 - t_1) \qquad (here, v < c)$$

This could be the proper calculation for the relativistic apparent length contraction in a case where moving reference frame is moving toward the source of the light or observed subject.

There would be a length elongation in case of the moving observer is moving away from the observed subject or the source of the light.

So, as I have cleared points above the time dilation equation describe in the Special theory of Relativity is wrong and there should be a time contraction when observer moves away from the observed thing. Similarly, there should be length elongation also if we consider observer specific constancy of the light true. If we consider Aether as an existence, then the Relativity in relation to Aether would be different From the Special theory of Relativity.

Photon clocks

Let me cover the topic of the photon clocks.

I have told ahead in this chapter that use of the photon clock example to show a time dilation is inappropriate. This is explained below.

There are two types of photon clocks described below. Photon clock type 1 is a classic photon clock used in the Special theory of Relativity where it shows example of the relativistic time dilation with simple calculations while photon clock type 2 is an assembly that enables us to derive a time contraction which is not described in the special theory of the relativity.

Photon clock type 1

Photon clock type 1 explains the time dilation concept in a simple manner.

(This chapter is here to understand an example of photon clock used widely in support of the special theory of relativity. This chapter is essential to understand photon clock type II explained ahead in book. This chapter is given from the perspective of special relativity theory. Author don't agree with all these.)

Here showing a figure of a photon clock type1.

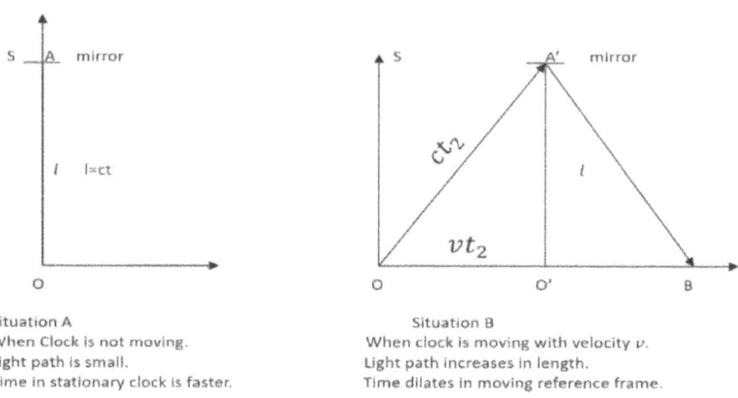

Figure 19.6

Simply from these figures, we can say that light path for the stationary observer increases when clock is moving in relation to stationary observer, speed of light c has to remains same. So, for compensation, time has to get slow down or has to be dilated

Please understand here that, t_2 is the dilated or slowed down time, means it is the time what clock takes to complete one half tick when moving. Means if time dilates then t_2 will be greater than t. While stationary clock will complete 1 tick for 1 second, there will be less than one second in moving clock.

Here,

As shown in Situation A, S is a stationary inertial reference frame. Clock is stationary with reference to S and also stationary in relation to observer. Light travels from O to A, strikes mirror at A and come back to O. Time to travel on one directional path for light is t.

Photon travels distance $l = OA$ in a time t.

Speed of light is c,

So,

$$l = ct \ \dots\dots\dots\dots\dots\dots\dots\dots\dots\dots\dots. (p.1)$$

Now,

As shown in Situation B.

S is stationary reference frame and photon clock is moving with velocity v along the positive X axis in reference to stationary observer. Now stationary observer from S reference frame will see the situation as shown in figure for situation B. Light travels from O to A', strikes mirror at A' and come back to B. Time in moving clock is t_2 in which photon travels OA' distance in situation B. Photon travels distance $OA' = ct_2$ in a time t_2. Clock moves a distance OO' in a time t_2 with the uniform rectilinear velocity v along with X axis. t_2 is a time that photon takes to move half of the path. Value of t_2 will always be greater than value of t. So, it can be said that the clock in moving reference frame will show less time compared to clock at rest.

So,

$$OO' = vt_2$$

$$OA' = ct_2$$

Applying Pythagoras theorem.

$$OA'^2 = OO'^2 + O'A'^2$$

$$c^2t_2{}^2 = v^2t_2{}^2 + l^2$$

Putting value of l from equation no (p.1)

$$c^2t_2{}^2 = v^2t_2{}^2 + l^2$$

$$c^2 t_2{}^2 = v^2 t_2{}^2 + c^2 t^2$$

$$t_2{}^2 (c^2 - v^2) = c^2 t^2$$

$$t_2{}^2 = \frac{c^2 t^2}{(c^2 - v^2)}$$

$$t_2{}^2 = \frac{t^2}{\dfrac{(c^2 - v^2)}{c^2}}$$

$$t_2{}^2 = \frac{t^2}{\left(1 - \dfrac{v^2}{c^2}\right)}$$

$$t_2 = \frac{t}{\sqrt{1 - \dfrac{v^2}{c^2}}}$$

As we know that $\dfrac{1}{\sqrt{1 - \dfrac{v^2}{c^2}}}$ is a factor Y,

$$t_2 = Yt.$$

Let me show an example to understand the dilation of the time.

If clock is moving with the velocity equal to $^c/_2$ means with the speed half of the speed of the light,
Then,

$$v = \frac{c}{2}$$

$$t_2 = \frac{t}{\sqrt{1 - \dfrac{(^c/_2)^2}{c^2}}}$$

$$t_2 = \frac{t}{\sqrt{1 - \dfrac{c^2}{4c^2}}}$$

$$t_2 = \frac{t}{\sqrt{1 - \frac{1}{4}}} = \frac{t}{\sqrt{0.75}} = 1.16t$$

$$t_2 = 1.16t$$

Moving clock will take more time to tick and it will be slower.

$$t = \frac{t_2}{1.16} = 0.86$$

This means for every second on the stationary clock moving clock will pass 0.86 seconds.

Photon clock Type-2

Now, I would like to show a photon clock type 2 that can show time contraction.

Photon clock Type-2 illustration.

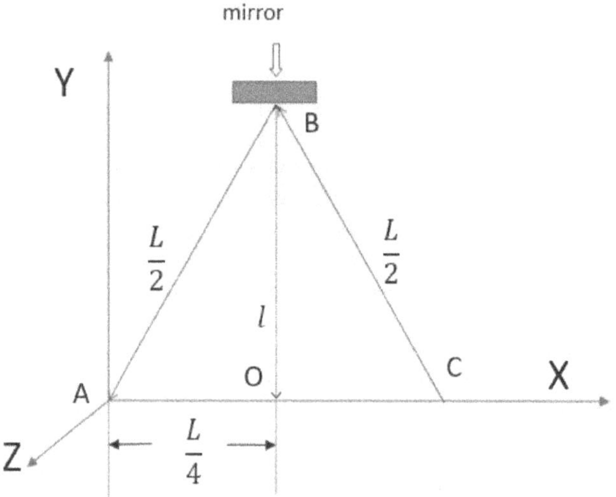

Figure 19.7

Here is the illustration of the photon clock type 2. Light source is at C, from where photon arises and travels toward B, where it gets reflected by mirror, and moves to reach to the point A. Total Distance travelled by light in a time t in nonmoving clock for stationary observer is CBA= $L = ct$. Angle BCA is 60°, angle BAC and angle ABC are also of the 60°. Angle BOA is right angle. So, Distances,

AB=BC=AC=ct/2=L/2.

Distance AO=1/2 *AC=L/4.

Height of the clock is l which is the straight minimum distance between X axis and mirror. Motion of light is two dimensional and on the path CBA.

Here, on the next page two reference frames are illustrated in a figure 19.8.

Reference frame *S* is a situation A

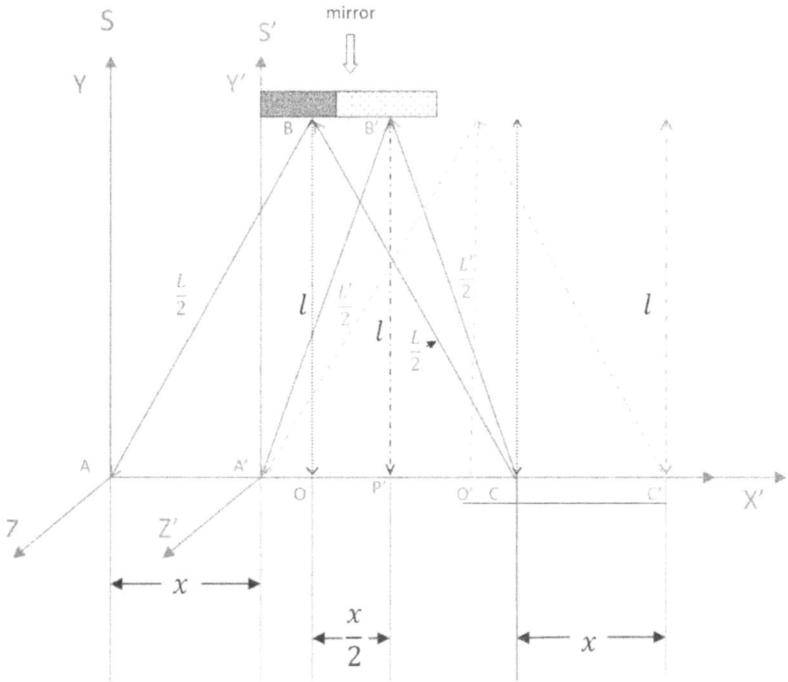

Figure 19.8

In a situation A, photon clock is not moving in reference to stationary observer in reference frame S. CBA is the path of photon, which photon completes in a time *t*. Length CBA is equal to L. $L = ct$. O is the point on *X* axis and BO is the minimum distance between *X* axis and *Y* axis. B is the point at where mirror is placed, and photon get reflected from this point at the height *l* towards the point A. OB= *l*.

So, here height of the photon clock is *l* which will remain same even when the clock will move in the case of situation B. The path of the photon makes a perfect triangle along with *X* axis, with each corner made up of 60°, So in a figure 19.7,

AB=BC=AC=2AO=2OC=L/2

So, AO=L/4

Light travels from C to B and then B to A. Total length of light path from C to A, via B is L, which light travels in a time t with the constant speed of light c.

So, Distance $CBA = L = ct$.

Distance $AO = L/4 = ct/4$

Let's find out the relation between l and ct.

In right triangle AOB, $\angle$AOB is the right angle.

So, according to Pythagoras theorem.

$$AB^2 = AO^2 + OB^2$$

$$\left(\frac{L}{2}\right)^2 = \left(\frac{L}{4}\right)^2 + l^2$$

$$\frac{L^2}{4} = \frac{L^2}{16} + l^2$$

$$\frac{L^2}{4} - \frac{L^2}{16} = l^2$$

$$l^2 = \frac{4L^2 - L^2}{16}$$

$$l^2 = \frac{3L^2}{16}$$

$$l = \frac{\sqrt{3}}{4}L$$

$$l = \frac{\sqrt{3}}{4}ct$$

$$ct = \frac{4l}{\sqrt{3}}$$

Now, Let's think about situations B.

Situations B.

In situation B, Photon clock with its reference frame S' moves with velocity v along with the Positive X axis. It travels a distance x in a time t with the velocity v along the Positive X axis. So,

$$x = vt = AA' = CC'.$$

This is the distance that clock moves in time t on positive X axis with the velocity v. Photon travels the distance L' in a time t_2 which is equal to ct_2 when speed of the photon is c. Here time is denoted as t_2 and not t, because photon has an observer specific constant speed, and here when clock moves, the path distance of photon for the stationary observer becomes short and to keep speed of the light same there would be a time contraction here which is reflected as a contracted time t_2. Here photon travels less distance for the stationary observer in a situation B compared to a distance travelled by photon in a situation A. You can see this path difference in the figure 19.8. You can see in a figure that base of the triangle formed by photon path in a moving clock appears contracted for the stationary observer as shown in in S' frame. Length $A'C$ is less than AC. The difference is equal to vt which is equal to $AA' = CC'$.

So, length of the base of the triangle in a situation B will be $A'C$, So,

$$A'C = AC - vt$$
$$A'C = \frac{ct}{2} - vt$$

So, in a situation B, height of the photon clock doesn't change, and it remains l.

Here, right triangle $A'B'P'$ formed with right angle $\llcorner A'P'B'$.

$A'P'$ is half of the $A'C$.

$$A'P' = \frac{1}{2}A'C = \frac{1}{2}\left(\frac{ct}{2} - vt\right)$$
$$A'P' = \frac{ct}{4} - \frac{vt}{2}$$

according to Pythagoras theorem.

$$(A'B')^2 = (A'P')^2 + (B'P')^2$$

$$\left(\frac{L'}{2}\right)^2 = \left(\frac{ct}{4} - \frac{vt}{2}\right)^2 + l^2$$

$$\left(\frac{ct_2}{2}\right)^2 = \left(\frac{ct}{4} - \frac{vt}{2}\right)^2 + l^2$$

Let's put the value of l^2.

$$l^2 = \frac{3c^2t^2}{16}$$

So,

$$\left(\frac{ct_2}{2}\right)^2 = \left(\frac{ct}{4} - \frac{vt}{2}\right)^2 + \frac{3c^2t^2}{16}$$

$$\frac{c^2t_2{}^2}{4} = \left(\frac{c^2t^2}{16} - \frac{ctvt}{4} + \frac{v^2t^2}{4}\right) + \frac{3c^2t^2}{16}$$

$$\frac{c^2t_2{}^2}{4} = \frac{1}{4}\left[\left(\frac{c^2t^2}{4} - \frac{ctvt}{1} + \frac{v^2t^2}{1}\right) + \frac{3c^2t^2}{4}\right]$$

$$c^2t_2{}^2 = \left(\frac{c^2t^2}{4} - \frac{ctvt}{1} + \frac{v^2t^2}{1}\right) + \frac{3c^2t^2}{4}$$

$$c^2t_2{}^2 = \frac{c^2t^2}{4} - ctvt + v^2t^2 + \frac{3c^2t^2}{4}$$

$$c^2t_2{}^2 = c^2t^2 - ctvt + v^2t^2$$

$$t_2{}^2 = t^2 - \frac{vt^2}{c} + \frac{v^2t^2}{c^2}$$

$$t_2{}^2 = t^2\left(1 - \frac{v}{c} + \frac{v^2}{c^2}\right)$$

$$t_2 = t\sqrt{\frac{v^2}{c^2} - \frac{v}{c} + 1} \quad \dots\dots\dots\dots\dots\dots\dots\dots\dots \text{(Q)}$$

(We neglected negative value for time here because time don't move backward here.)

This is the relationship between variated time t_2 with time t in stationary clock.

Here path contracts, So, time will also contract for compensation to keep the speed of light c constant when clock is moving with velocity v.

So, for each second in a stationary clock, time passing in a moving clock would be more than a second, means moving clock will tick faster. Means moving clock will take less time to complete a single tick.

So, here t_2 will be smaller than t in contrast to Type-1 Photon clock, where time dilates as shown in a topic just before this topic.

t_2 is a time taken to complete 1/2 tick. When it is smaller, clock shows more time and when it is bigger clock shows less time.

This was a calculation according to the Special theory of Relativity and this prediction is not mentioned anywhere in the special theory of the relativity.

Here the point is, if there is a moving inertial reference frame with a light moving with its source in that reference frame and there is a stationary observer in a stationary inertial reference frame then it is not always necessary that there will be an elongated path of that light for the stationary observer compared to path observed by observer in moving reference frame. reverse is also possible as shown in this Photon clock type 2 example. Assembly of the photon clock is a theoretical and we just need a half tick so reflecting back of the light photon for further tick is not required here, but it can be achieved by putting vertical mirror glass at a point A and putting another mirror below at a distance l which should be aligned in a line and parallel to the mirror at B to reflect light back again at the source. So, according to the relative direction of the light beam there can be a variation on the length of the observed path by stationary observer when source of the light is moving in relation to that observer. Application of the single time dilation equation is not possible here for all possibilities depending on the direction of the light beam. So, example of the photon clock to show that time dilation is wrong.

Relativity in relation to Aether.

Second Postulate of the special theory of Relativity *"Speed of light is constant for the observer irrespective of the reference frames and relative speed of the source of light."* Can not be taken as it is in relation to Aether-Brahm theory, but it should be transformed as *"speed of the passage of the time depends on the relative motion of the mass in relation to aether and also depends on the relative balance of the flux causing push effect."* This explains time dilation in a moving inertial reference frame and also a time dilation in a gravitational field. The second postulate of the special theory of Relativity *"Laws of physics are same and can be stated in their simplest form in all inertial frames of reference"* is not competent with this theory because inertial reference frame has modified laws according to its motion in relation to aether. For example, if we imagine a reference frame moving with the speed of aethereal linear wave, then there could not be a possibility of matter within that reference frame which is a different law than a reference frame moving slower in relation to Aether where existence of matter is possible. So, the world doesn't remain same to subjects in all kinds of motions. Time dilation is neither a bilateral relation nor dependent on relative motions of masses with each other, it just depends on the relative linear motion of the matter in relation to Aether-Brahm and flux balance. One also has to think about relative momentum in relation to Aether and Brahm only.

Passage of time:

We have defined a primary matter as a spinning vortex of waves in Aether. Here time is a change that appears due to spin of the vortex. When there is a delay in a time to complete the spin due to displacing motion for the vortex in relation to Aether it would slowdowns the passage of time for that vortex.

If we place a clock steady in relation to aether-matrix and there is a clock moving in relation to that steady clock, then time in a moving clock will be slowed down. Amount of such a slowing down would be dependent on the relative motion in relation to aether-matrix. If hypothetically the clock would move with the speed of the light the time would be standstill for it due to no spins of the primary masses. So, in reality as much the speed in relation to Aether there is a slower passage of time. Here relative speed more than a speed of light is possible because there can be a speed more than the half of the speed of light in one direction. When two material objects moving on opposite direction with each other, then they are moving apart with the speed more than that of light. It would be perceived as a doppler effect + wave transformation if those two objects are at very far distance from each other. So, Relative speed more than the speed of light is possible in the context of Aether-Brahm theory but that can exceed or reach to 2c, double the speed of light.

Such objects which have speed more than half of the speed of light and opposite direction of the motion would not become invisible for each other and light from one object would reach to other object. This is because once light enter Aether it moves with a constant speed in relation to Aether in a vacuum. So, eventually it would reach to the other object. There is a relative slowing down of speed would be experienced by one moving subject for another moving subject due to time dilation but that would not change the absolute situation described above. When can an object become invisible for another object? Means a light emitted by the first object can never reach to the second object due to their motions. Such situation is never possible if there is a no obstruction on the way of the emitted light because once light is emitted in the Aether, it would move with the speed equal to c in relation to Aether and no material object could surpass this speed in relation to Aether. Light will eventually catch the second object.

The direct experience of fastest observed speed can be that of the speed of light only if the space is not being added in between the path.

Let's calculate time dilation when the "m" vortex is moving in relation to aether with the speed v.

We know that if it is hypothetically forced to move with the speed c than it will lose its spin and it will become waves moving with the speed of light and there won't be any time passing for that ex-vortex because it is now no longer a vortex or a mass.

If dilated time is a t'

Then,

$$t' = t_0 \frac{c - v}{c}$$

Where t_0 is a non-dilated time or a fastest time.

This means when linear motion in relation to Aether is zero, then

$$t' = t_0 \frac{c - v}{c} = t_0 \frac{c - 0}{c}$$

$$t' = t_0$$

Means there is a no time dilation.

When linear motion will be equal to c,

$$t' = t_0 \frac{c - c}{c} = t_0 \frac{0}{c}$$

$$t' = 0$$

Means mass will be annihilate and no time applicable for massless Aethereal waves which travels at a speed c.

Let's calculate time dilation for the object moving with speed c/2.

$$t' = t_0 \frac{c - v}{c} = t_0 \frac{c - c/2}{c} = t_0 \frac{\frac{2c - c}{2}}{c} = t_0 \frac{c/2}{c} = \frac{t_0}{2} = (0.5)t_0$$

Means time will be slowed down to the half of its rate.

What could be the time dilation according to special relativity here?

$$t' = \frac{t - \frac{vx}{c^2}}{\sqrt{1 - \frac{v^2}{c^2}}}$$

$$t' = \frac{t - \frac{vct}{c * c}}{\frac{\sqrt{c^2 - v^2}}{c}}$$

$$t' = \frac{t - \frac{vt}{c}}{\frac{\sqrt{c^2 - v^2}}{c}} = \frac{\frac{ct - vt}{c}}{\frac{\sqrt{c^2 - v^2}}{c}} = \frac{ct - vt}{\sqrt{c^2 - v^2}} = t\frac{c - v}{\sqrt{c^2 - v^2}}$$

Putting $v = c/2$

$$t' = t\frac{c - \frac{c}{2}}{\sqrt{c^2 - (\frac{c}{2})^2}} = t\frac{\frac{c}{2}}{\frac{\sqrt{4c^2 - c^2}}{2}} = t\frac{c}{\sqrt{4c^2 - c^2}} = t\frac{c}{\sqrt{3c^2}} = t\frac{1}{\sqrt{3}}$$

$$t' = (0.57)\, t$$

This value is near to the value calculated ahead.

Derivation of aethereal relativistic transformation equations

If there is an observer specific constancy of the speed of light, then there could be following possible calculations that would help in the calculation of the time dilation due to motion of the mass in relation to Aether. That is applicable to the non-spinning cluster of mass moving with a constant rectilinear velocity in a single direction or in any directions.

There is a figure 19.2, shown below to explain time dilation and co-ordinates transfer.

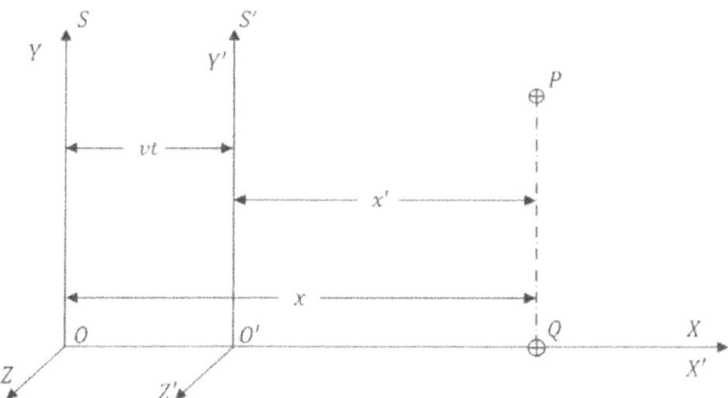

Figure: 19.2

Here two inertial reference frames S and S' with attached cartesian planes are shown which defines co-ordinates of the point P as (x, y, z, t) and (x', y', z', t') respectively in S and S'. S represents Aether matrix. S' inertial reference frame is moving with uniform rectilinear velocity "v" along the direction of positive X axis of S frame which contain an observer or a clock. There is no movement of S' along Y or Z axis of the frame S. Both frames are aligned in a way that O and O', co-insides at time $t = t' = 0$. P is the point selected randomly in the stationary reference frame S toward Positive X axis side of S. Q is the reference point on X axis in the reference frame S with co-ordinates $(x, 0, 0, t)$. In a time "t" photon travels perpendicular to y axis from the point P for a distance equal to $OQ = x = ct$. and meanwhile during the same time t, S' frame moves a distance vt along positive X axis.

Here there are two observers O and O'.

O is steady in relation to aether. So, it would show fastest time.

O' is moving in relation to aether, so in relation to O also.

That's why time would be slower at O'.

We accept observer specific constancy of speed of light here for now. only one postulate of the special relativity.

Now,

x is the co-ordinate of P on the X axis.

$$x = ct \qquad \text{...... (19.1)}$$

And

$$x' = ct' = ct - vt \qquad \text{...... (19.2)}$$

$$ct' = ct - vt$$

$$t' = \frac{ct - vt}{c}$$

$$t' = t\frac{c - v}{c}$$

This is the same equation we have assumed ahead in this chapter.

$$x' = ct' \qquad \text{....... .... (19.3)}$$

Now put the value of x' & t' in this equation (19.3)

$$x' = ct'$$

$$x' = c.t\frac{c - v}{c}$$

$$x' = x\frac{c - v}{c}$$

Here $v < c$.

So, aethereal relativistic transformation equations are,

$$x' = x\frac{c - v}{c}$$

$$y' = y$$

$$z' = z$$

$$t' = t\frac{c - v}{c}$$

Inverse aethereal relativistic transformation equations would be,

$$x = \frac{ct'}{c - v}$$

$$y = y'$$

$$z = z'$$

$$t = t'\frac{c}{c - v}$$

Time dilation:

Dilated time *is* t'

$$t' = t\frac{c - v}{c}$$

In this equation there is a no dependency on x.

x was there, just for calculations. Its value must be more than 0 to preserve nonzero value of t and v.

This above equation shows dilated time. According to it.
For $v = 0$, $t' = t$
For $v = c$, $t' = 0$
Means no time dilation for steady object in relation to Aether and time is not passing at a speed c.

Time dilation would be,

$$\Delta t = t - t'$$

$$\Delta t = t - t\frac{c - v}{c}$$

$$\Delta t = \frac{ct - ct + vt}{c}$$

$$\Delta t = \frac{vt}{c}$$

$$\Delta t = t\frac{v}{c}$$

With the speed equal to v in relation to Aether there would be $\frac{v}{c}$ amount of time dilation per unit of time.

This is a time dilation for a relative speed in relation to Aether-matrix and not in relation to each other or one another.

If this does not match with the observed data than it means that there is a no change in the rate of spin when vortex is dragged in Aether by unequal flux gradient. What we observe as a time dilation is due to distortion of the shape of the mass occurred in an environment of unequal flux pressure that may lead to affect the behaviours and oscillation of the atoms and crystals.

Here the situation in a diagram is plotted to understand motion in relation to Aether matrix. Because Aether matrix is pervaded everywhere there could not be a negative velocity for a motion in relation to Aether so phenomena like a time contraction is not applicable here which is described in a criticism of relativity for time dilation in a relative motion of the objects with each other.

When object moves in relation to Aether its time slowed down. When object is steady in relation to Aether it has fastest moving time. Tides in a Brahm creates reference frames where the laws of the physics appeared to be same due to same amount of the inertia. when that reference frame would achieve a speed near to speed of light then there would be a different law of physics there because then a slight motion in a direction of the motion of that reference frame would annihilate mass and motion in opposite direction of the motion of that reference frame would preserve mass from annihilation. At a speed of c for the reference frame there would not be any mass possible in that reference frame. So inertial reference frame can't hold law of the physics as it is in all reference frame. It is our inability to detect a change made by inertia forming tides in mass that hides change of law of physics in moving inertial reference frame in relation to Aether. Flux in a Brahm can hold integrity of the mass in a relative motion in relation to aether to some extent only but not at a speed near or equal to c.

Do tides or flux in a Brahm affects the rate of time passage on the object? If there is a time dilation in a gravitational field than it would be either due to aether compression or curvature or due to unequal pushing flux or may be due to both.

Suppose the passing Time is a one tic when primary matter vortex spin one circle. When vortex has no linear motion in relation to aether there is a no unequal pushing flux, it spins with maximum speed and time passes with maximum speed. When vortex moves in relation to aether with linear or random movements other than its Spin, then there would be an inequality in a strength and amount of the pressure forming flux and its linear or random movements occurs with the cost of circular motion. So, it takes more time to complete one circle or one Spin when it is moving in relation to aether. Why is this so? This is because matter is ultimately formed by waves in Aether and there could be a constant finite speed of wave propagation in a medium for waves forming primary vortices of matter, so when circular waves (vortex) move in a linear direction then its speed in a circular direction decreases with increase in its linear speed.

This also explains why speed of matter equal to or more than a speed of the linear wave in Aether can't be possible. This is because linear waves have no circular motion or spin, so it travels with maximum possible speed. If matter has to travel with the speed of aethereal linear wave then it has to lose spins of all fundamental primary mass vortices and have to travel as a cluster of linear waves, and in that situation, it will no longer remain a matter, but it will become aethereal waves.

Explanation of the observations and predictions of the theory of Relativity with Aether-Brahm theory.

(I) Gravitational Lensing.

Two possibility explains this.

1. Gravity affects light because light is a material thing. Flux that causes push gravity also pushes light photon and thus gravitational lensing works and is not due to curvature in spacetime.
2. Aethereal units constricts and dilates according to gravitational push and mass. Thus, arrangement of aethereal units in and around mass creates geodesics which direct the path of the light wave.

So, Aether-Brahm theory can harbour the characteristic of the gravitational lensing still it doesn't determine the nature of the light.

(II) Blackholes, Neutron stars and singularity.

Because this theory accepted push gravity it doesn't allow limitless gravity. This limitation of gravitational push explains quantum nature of the mass. It explains weak force. It explains supernova. Thus, limitless vicious gravity has no role in Aether-Brahm theory and thus relativistic Blackhole has no place in his theory, but a highly condensed mass could be there within a limit of certain amount of mass create due to direct pressure of outer mass due to push on that outer mass. Means Though Gravity is weak inside the core of Star or at the centre of the Galaxy, Pressure is higher there due to direct push of outer material. If such pressure crushes an Elemental mass than it be converted to more condensed mass made up with subatomic particles and smaller units. But that is not a singularity.

(III) Worm holes:

No relativistic black holes so no relativistic Wormholes.

But if we are talking about a condensed areas within Aether then it is possible.

(IV) Time Travel:

If we are talking about time ravel with slowing down of time with help of time dilation at high speed or in extreme Gravity, then it is possible.

20. Brahm without matrix substance Aether layer

As described in the introduction at the beginning of the book, this chapter provides a model that don't uncloses substance-aether-medium and describes only Brahm without Aether to represent all existences including light with dual nature.

If this is possible to describe world without an Aether than why Aether was included ahead in this book? It is done because possibility of that can't be ruled out. But still Aether is not essential to describe the world with Brahm theory.

So, here is a Brahm without Aether that can explain whole world without Aether.

As described in a chapter Honi, multiple infinite blasts form an environment, that is a gravitational environment where different sized corpuscles move in all directions and forms structures and clusters which further forms hollows, tori, and intermediate structures, that results in the formation of the material or matter. Here we are not accepting a matrix substance formation at all. Here only moving flux of masses and primary masses forms everything. Here waves in Brahm are not possible and what we called EM waves should be explained in alternate ways. They could be moving particles. What provide them a wave Nature? It is their interaction with flux while motion. Like, when a charged particles moves in a magnate field, it has a helical path. Light or other EM waves can be a fish projectile like structure also which have a wave Nature due to their propulsive movements of the moving parts. This are also subjected to tapering phenomena and explain cosmological redshifts. This also explains bending of the light around heavy massive objects. Continuous blast of small materials or annihilating previously formed material provides an emergent space-matrix with gravitational effect and dark energy.

These EM waves can also be a Torus particle which have oscillation of its axis while motion which creates a particular frequencies and wave lengths. There are also subjected to tapering.

Rest of the things remains same as a Brahm model, including infinity without including Aether like matrix material forming scales.

So, this can also be a model including everything where primary masses remain undefined solids.

Here Flux is a dynamic thing that can provide a relativistic symmetry, and relativity can be aligned well in this environment lacking Aether as a universal reference frame. Here the cause of the Time dilation can be related with inequality of Flux. Imbalance of Flux around the mass leads to two results, 1. Motion 2. Time dilation.

Motion is explained in the chapter of Inertia can be applied here. Time dilation is due to longer path to complete a spin for a primary mass moving in orbit of the Torus. Here reference for the relative motion is a reference frame where Flux is balanced from all side.

21. Substance Aether in a liquid state.

The possibility of Aether substance in a liquid state is also there. The process of the formation of the Aether at various scale is show in the chapter 2, "Honi". This substance Aether is described in the Elastic state. The possibility of Elastic fluid like state is also there. The state of Aether depends on the nature of the units of Aether. Prediction of all characteristics of Aether unit is not possible with insights generated with this theory of Brahm.

What if the state of Aether is liquid like and not a solid like? These will make internal Aether currents possible. The matter formation as vortices of waves in Aether remains same, propagation of aethereal waves remains same. Torus formation would be same. But internal current would add additional properties like Aether drag with mass. Aether motion may be there in torus. Aethereal currents may explain the following observations.

1. Observation of the Great attractor has amazed scientists. There is no satisfactory explanation of that observation. If it is explained with aethereal current that it is understandable as a drag of galaxies along with aethereal flow. Because mass is described as made up of aethereal waves, it must be dragged along with aethereal flow.
2. Inertia can be explained in a better way. If galaxies are dragging and carrying aether medium with them and solar system is also dragging aether medium with it then explanation of inertia becomes simple and easy with addition of this drag.
3. Observations of solitary stars moving in the orbit around nothing observable which is believed to be a hidden black hole can be explained as a vortex or whirlpool in aether liquid dragging a star along with it.

There would be some conflicts also in excepting aether liquid. Those conflicts would be same for both liquid and solid aether, so this is not a new and special drawback.

Chapter Honi provides a logic up to formation of the gravitational environment. Then multiple possibilities follow the path ahead. Brahm with scaled Aether, Brahm without Aether, Brahm as solid scaled Aether, Brahm as liquid scaled Aether, all these possibilities are explained in this book. More observations of objects in the existence can help to understand if Aether is there or not and if it is there that those observations could help in understanding the state of the Aether. Observations of, galactic centres, streams in observable cosmos may provide some clues.

The Michaelson Morely experiment could be more sensitive if a vertical beam would be there in the experiment. There may be an aether wind after some amount of dragged aether for some height. All experiments were carried out on the surface of the earth only. Vertical beam perpendicular to the surface of the Earth can catch drag effect more efficiently. Making such instrument is costly and time-consuming procedure.

22. Kinetic Theory of Gravity, its history and criticism

Kinetic theory of gravitation by Nicolas Fatio de Duillier

First known scientist who gave the well detailed explained kinetic theory of gravity was Nicholas Fatio de Duillier. Shadow gravity theory of Nicolas Fatio de Duillier:

This theory is probably the best-known mechanical explanation.

It was first developed by **Nicolas Fatio de Duillier** in 1690. Later, it was rediscovered by others, including **Georges-Louis Le Sage** (1748), **Lord Kelvin** (1872), and **Hendrik Lorentz** (1900). However, it faced criticism from **James Clerk Maxwell** (1875) and **Henri Poincaré** (1908). Nicolas Fatio de Duillier (1664–1753) was a Swiss mathematician, natural philosopher, and a close associate of Isaac Newton.

Nicolas Fatio de Duillier

He developed one of the earliest and most detailed kinetic theory of gravitation.

Fatio's work on gravity was primarily articulated in his unpublished manuscript, *De la Cause de la Pesanteur* (On the Cause of Gravity), composed around 1690. While never formally published in his lifetime, its ideas were circulated in correspondence and discussed among prominent scientists of the era, including Christiaan Huygens and Isaac Newton.

Here's a breakdown of Fatio's kinetic theory of gravitation:

The Core Idea: Push-Shadow Gravity

Fatio's theory is a classic example of **"pushing gravity"** or **"shadow gravity"**. It sought to explain the apparent attraction between massive bodies not as an inherent, mysterious pull, but as the result of a net *push* from an external, unseen medium.

1. The Aethereal Sea of Corpuscles: Fatio proposed that the universe is permeated by an immense number of tiny, invisible, and perfectly hard (or nearly so) particles, which he called "inframundane particles." These particles move randomly and at incredibly high speeds in all directions throughout space.

2. Omnidirectional Bombardment: An isolated object in this cosmic shower would be bombarded equally from all sides by these gravific corpuscles. The impacts would exert pressure on the object, but because the forces are balanced, there would be no net directional movement.

3. The Shadowing Effect: The crucial insight comes when two or more massive bodies are present. Each body would partially shield the other from the incoming stream of gravific corpuscles. Imagine two bowling balls in a hailstorm. Each ball would create a "shadow" for the other, meaning fewer hailstones would strike the inner, facing sides of the balls compared to their outer, unshielded sides.

4. Net Inward Push (Gravity): This shielding effect creates an imbalance of pressure. The outer sides of the bodies, being fully exposed to the corpuscle bombardment, would experience a stronger push than the inner, shielded sides. This differential pressure results in a net force that *pushes* the two bodies towards each other, mimicking gravitational attraction.

Key Assumptions and Challenges:

To make this theory work and align with observed gravitational phenomena, Fatio had to make several audacious assumptions:

- **Extreme Penetrability:** For gravity to be proportional to *mass* (and not just surface area), Fatio had to postulate that ordinary matter is almost entirely transparent to these gravific corpuscles. He argued that even seemingly dense materials like gold must consist mostly of empty space, allowing the tiny aether particles to pass through with minimal interaction. This was a radical idea for his time.

- **Inelastic Collisions (or Reduced Speed):** A significant challenge was explaining how momentum is transferred without causing an immediate and continuous acceleration. If the corpuscles were perfectly elastic and rebounded with their original speed, they would simply bounce off without imparting a net force. Fatio wrestled with this, suggesting that the collisions were either slightly inelastic, or that the reflected particles lost some speed or gained internal vibrations, thus exerting less momentum upon reflection. This would create the necessary net inward push.

- **High Speed, Small Size, and Rare Collisions:** To avoid any observable drag on celestial bodies moving through this aether, the inframundane particles had to be incredibly small and move at extremely high speeds. Furthermore, to prevent the aether itself from dissipating its energy through self-collisions, Fatio assumed their diameters were minuscule compared to their mutual distances, making inter-corpuscle collisions very rare.

- **Inverse-Square Law:** Fatio realized that the shielding effect, given the geometry of spheres and the flow of particles, could naturally produce an inverse-square law of attraction, consistent with Newton's observations.

Why it didn't prevail:

Despite its elegance and the intellectual effort Fatio poured into it, his kinetic theory of gravity (and its later iterations by Le Sage and others) ultimately failed to gain widespread acceptance and was superseded for several reasons:

- **Thermodynamic Inconsistencies:** The theory implied an enormous amount of energy constantly being transferred from the gravific corpuscles to matter, which should lead to continuous heating of all objects in the universe. This "thermal death" prediction was not observed.

- **Predicting Drag:** Even with very high speeds, the constant bombardment should cause a drag on celestial bodies, slowly decelerating them and causing orbits to decay. This was inconsistent with astronomical observations.

- **No Empirical Evidence:** There was no direct experimental evidence for the existence of these gravific corpuscles or the aether they inhabited.

- **Newton's Success:** Newton's *Principia* provided a highly accurate and predictive mathematical framework for gravity, even if it didn't explain *how,* it worked mechanistically. It was a tremendously successful descriptive theory.

- **Einstein's General Relativity:** Ultimately, the theory of General Relativity in the 20th century provided a completely different, geometrically based understanding of gravity as the curvature of spacetime, which was powerfully supported by experimental evidence and rendered the need for a gravitational aether obsolete.

Other scientists with same or similar theory:

Robert Hooke & James Challis:

Reverse to the kinetic theory proposed by Nicholas Fatio De Duillier, In 1671, Robert Hooke even before that proposed a theory of gravitation with a mechanism for direct attraction. He suggested that all bodies emit waves in every direction through the aether. He believed other objects, interacting with these waves, would then move toward the wave source. Hooke drew a parallel to how small items on rippling water drift towards the disturbance's centre. Later, between 1859 and 1876, James Challis developed a similar mathematical theory. Challis's calculations showed attraction occurred if the wavelength was long compared to the distance between gravitating bodies, while repulsion happened with short wavelengths. He even attempted to explain all other forces by combining these effects. However, the theory faced criticism; Maxwell pointed out it would necessitate a continuous, infinite expenditure of energy to produce these waves. Challis himself acknowledged the difficulty in reaching a conclusive outcome due to the processes' intricate nature.

Lord Kelvin and Carl Anton Bjerknes:

Pulsation

In 1871, both Lord Kelvin and Carl Anton Bjerknes proposed a theory where all bodies pulsate within the aether. They drew an analogy to how two spheres pulsating in phase within a fluid will attract, while out-of-phase pulsation leads to repulsion. This same mechanism was also applied to explain the nature of electric charges. Notable scientists like George Gabriel Stokes and Woldemar Voigt also investigated this hypothesis. However, this theory faced significant criticism. To account for universal gravitation, one would have to assume that all pulsations throughout the universe are perfectly in phase, which seems highly unlikely. Furthermore, the aether would need to be incompressible to ensure attraction over larger distances.

Maxwell also pointed out that this process would require continuous creation and destruction of the aether."

Pierre Varignon (1690):

Gravity by Unequal Pushes

Varignon's contribution in 1690 presents a compelling, if ultimately speculative, mechanical explanation for gravity. He envisioned the universe filled with **aether particles** constantly bombarding all bodies from every direction. The key to his theory was the introduction of a "limitation boundary" near the Earth's surface. According to Varignon, these aether particles could not penetrate beyond this boundary. This critical assumption led to a fascinating conclusion: if a body was situated closer to the Earth than to this boundary, it would experience a greater flux of aether particles (and thus a stronger pushing force) from above, where the aether was unrestricted, compared to from below, where the Earth acted as a "shield" or the boundary prevented penetration. This differential pressure, a net downward push, was his proposed mechanism for gravitational fall.

Mikhail Lomonosov (1748):

Aether's Proportionality and Permeability

Mikhail Lomonosov, a renowned Russian polymath, also delved into aether theories, drawing inspiration from predecessors like Christiaan Huygens and Nicolas Fatio de Duillier. In 1748, Lomonosov proposed that the effect of the aether on matter was proportional to the *complete surface of the elementary components* from which matter is constituted. This suggests a very fine-grained interaction, where the aether somehow interacts with the most fundamental parts of an object. He further assumed an "enormous penetrability" of bodies by the aether, which was crucial for explaining why gravity affects the entire mass of an object, rather than just its external surface. Without this high penetrability, larger objects would simply "shield" themselves more effectively from the aether, leading to a different gravitational behaviour than observed. Lomonosov did not provide a clear and mathematically rigorous description of *how* this aether interaction precisely led to the observed law of gravitation.

John Herapath (1821):

Kinetic theory and the Heated Aether

John Herapath's attempt in 1821 is particularly interesting because it sought to apply the burgeoning **kinetic theory of gases**—a successful framework for understanding the behaviour of gases in terms of the motion and collisions of their constituent particles—to the phenomenon of gravity. Herapath hypothesized that bodies, through their presence, heated the surrounding aether. This heating would cause the aether to expand and lose density in the vicinity of the mass. The consequence? Other bodies would then be pushed towards these regions of lower aether density, similar to how a less dense fluid would be displaced by a denser one. This created an apparent "attraction" based on a pressure differential within the aether.

Gabriel Cramer & Redeker

Swiss mathematician Gabriel Cramer in 1731 published a decertation containing sketch and theory very similar to Fatio's theory. In 1736 the German physician Franz Albert Redeker also published a similar theory.

Le Sage:

In 1748 Georges Louis Le Sage submitted the same theory to the Academy of Sciences at Paris, which was rejected and not published due to similarity with the theories by Nicholas Fatio, Cramer and Redeker. According to Zehe Le Sage had Fatio's papers in his possession and Le Sage contributed nothing essentially new and he often did not reach Fatio's level.

Peter Guthrie Tait called the Kinetic theory of Le Sage the only plausible explanation of gravitation which had been propounded at that time. went on by saying:

"The most singular thing about it is

that, if it be true, it will probably

lead us to regard all kinds of

energy as ultimately Kinetic."

James Clerk Maxwell's review of Kelvin-Lesage theory was published in the Ninth Edition of the Encyclopaedia Britanica under the title Atom in 1875. He wrote,

"Here, then, seems to be a path

Leading towards an explanation of

The law of gravitation, which, if it

can be shown to be in other

respects consistent with facts,

may turn out to be a royal road

into the very arcana of science."

He also wrote that this kinetic theory is the only one theory of gravitation developed so far which is capable of being attacked and defended.

Georg Christoph Lichtenberg believed that every explanation of natural phenomena must be based on rectilinear motion and impulsion, and Le Sage's theory fulfilled these conditions. He wrote Le Sage's theory embraces all of our knowledge and makes any further dreaming on that topic useless. He went on by saying: "if it is a dream, it is the greatest and the most magnificent which was ever dreamed..."

According to me, all credits go to Nicholas Fatio for introducing such a mechanism for the very first time in known history of science. This is also because it was known to Fatio that Cramer had access to a copy of his main paper, so he accused Cramer of only repeating his theory without understanding it. It was also Cramer who informed Le Sage about Fatio's theory in 1749.

Keller and Boisbaudran:

In 1863 Francois Antoine Edouard and Em. Keller presented a theory with mechanisms like Le Sage but replacing particles with aethereal longitudinal waves. in 1869 Paul-Emile Lecoq de Boisbaudran proposed the same theory as Keller.

Many other scientists showed their views and either supported or criticized the kinetic theory of the Gravity.

The Decline of Aether Theories for Gravity

Despite the intellectual ingenuity of these and other aether theories for gravity, they ultimately faced insurmountable challenges.

1. **Lack of Empirical Evidence:** No experiment could definitively prove the existence of a gravitational aether, let alone measure its properties.

2. **Inconsistencies with Observation:** Many aether models struggled to quantitatively reproduce Newton's inverse-square law of gravity without introducing ad-hoc assumptions.

3. **Thermodynamic Problems:** As seen with Herapath, many kinetic or pushing aether models predicted an enormous amount of energy dissipation and heating, which was simply not observed in the universe. Planets would quickly lose energy and spiral into their stars.

4. **Rise of Alternative Theories:** The greatest blow came with the advent of more robust and empirically supported theories. Isaac Newton's law of universal gravitation, while not explaining *why* gravity worked, accurately described *how* it worked, predicting planetary motion with remarkable precision. Later, Albert Einstein's theory of General Relativity in the early 20th century provided a revolutionary new framework, explaining gravity not as a force mediated by a substance, but as a manifestation of the curvature of spacetime itself caused by mass and energy. This theory rendered the gravitational aether superfluous. Various scientists represent vital steps in humanity's long quest to understand the fundamental forces of the universe. The legacy of the aether, even in its eventual dismissal, underscores the deep human desire to understand the underlying mechanisms of the cosmos.

Criticism:

Gravitational shielding:

In a kinetic theory of Gravity mass produces a shadow effect. When much mass accumulates such shadows become overlapped. Though mass was imagined as extremely porous in a theory by Fatio, the shielding effect due to shadow overlapping must be there which results in the violation of the equivalence principle and render the theory inconsistence with both Newtonian theory and General relativity.

In this theory of Brahm this gravitational shielding is accepted and is described as a cause of quantization of maximum mass sizes at particular density scale. This effect defines subatomic particle sizes and also restricts maximum star size and maximum galaxy size. This shielding effect doesn't allow the formation of relativistic blackholes with singularity. There is nothing other than this effect which can explain these size limits plausibly.

Drawbacks proposed by the science community for Fatio's Kinetic theory of Gravity:

1. That theory demands the extremely porous nature of the mass (now it is a proven fact.)
2. That theory demands a very abundant flow of inframundane particles or similar things causing impulsion. (The theory of Brahm is based on such environment derived in the thought experiment in the chapter "Honi" which is described ultimately as aethereal waves)
3. Collision of the flux with mass must be inelastic and pure elastic collision don't allow a push gravity to work.

This is the toughest part of the kinetic theory which must be explained. Objection by Huygens put Fatio in dilemma for at least three years.

Explanation is only one which can be described as inelastic collision with scattering after fragmentation and heat generation. Here, such heat generation becomes so intense that that could incinerate all matter within a few seconds. This assumption was there because of the poor understanding about the matter. Matter has condensed mass at extremely small nucleus and a shield of field surrounds nucleus which forms the most area of the mass. Temperature at nucleus can't affect the gross temperature of the mass. As we have assumed hollows like stars so they must have a high temperature at their level.

The query of fundamental cause of the Inertia actually brings this problem. Inertia is explained in this book in relevant chapter, but the explanation remains incomplete due to inter-dependency of the inertia of mass and inertia of tides in the aether. In a chapter "Honi" Primary acceptance of law of inertia is there when we have assumed motions of masses in space. Then subsequently Aether was degenerated in gravitational environment. Waves in that Aether obeys law of inertia and wave dynamics explains the whole existence and laws of physics.

The answer to the question, why all these follows the law of inertia can be given as "this is because the structure and distribution of the material Aether is such that a primary law of an existence "mass acquires a space" eventually reflects as the law of motions and law of inertia. Tides can be settled down in aether as a result of negative interference with opposite tide by converting kinetic energy to finer form or in a heat form. I would like to give an example. If we push a transversely placed rod with the tip of the finger, then it doesn't move just a touched part of the rod, but it moves the whole rod. What gives momentum to the rest part of the rod if tides of the finger are applicable to the little portion of the rod only? Here tide transfer doesn't happen directly with same tide donation. Aether plays a role. Obstruction leads the extra tide of the finger to get settled in aether and some part to reflect back which balances motion cessation of the finger. The direct interaction of the fields of the material of the finger with field of the material of the rod moves a whole aethereal wave complex, a rod and this process it forms new extra tides in the direction of the motion of the rod. It causes imbalance in the tides of the rod that leads to inertia and continues motion that doesn't stops until stopped by something.

What is the nature of collision of flux with mass in Brahm theory? Vast size deference between flux units and primary mass is the key. If we see the surface of the sun, we can see that there is very high temperature at the sun surface but still it doesn't evaporate the sun in few seconds. The cause of the heat production in the sun is considered as a fusion rection. But the force that powers this reaction is gravity. So, after the collision net force applies the pressure on the mass toward other mass and that may result in heat, fusion, collapse or chemical reaction. So, kinetic energy is not destroyed but it is being converted to other forms and behaves accordingly. Sun radiates light and other radiation, but they don't compensate in opposite direction of the gravity forming flux moving toward the sun, so, effect of gravitational push can't be cancelled. Similar things happen to the Torus and Hollow structure of the quantum sizes. Does flux reflect back? As explain in the description of the Perihelion of the mercury, it is described that spin of the sun causes drag of the reflecting flux around it that changes speed of the mercury and thus orbit becomes acentric and Rosset pattern type. The amount of the reflected flux is lower than the gravity forming flux. Thus, gradient is created. Does this phenomenon eventually vanish the total amount of flux in the universe? I don't think so, because the only origin of the flux is not its existence since the formation of the gravitational environment. I have mentioned ahead that there are multiple sources that continuously forms flux. Quasar jets, Gamma radiation from pulsar, magnetar, neutron stars etc. There may be a large source also. As assumed in the chapter of Aethereal structures, there is a description of the Structure Brahmand which is an egg like globe of the universe. If it behaves as the unit of the higher aether, then there must be vibration and motions of that units occur in that environment. That motions and vibration can generate tremendous amount of the flux as waves from the shell of this Brahmand. It must be the continuous process. This framework allows so many imaginations in support of the generation of the flux. The Brahmand must have a buffer range within which it harbours energy and extra energy must be radiated by it in its outer environment and thus the law of conservation gets the condolence. So, there is a no reason to discard push gravity model by imagining impossible nature of collision that can generate the push or imagining that a loss of kinetic energy can violates the conservation law. Too much imagination is not advised in absence of necessary data because that lead to overthinking and paranoid thoughts as described in the chapter 0.

Citations

For the topics mentioned below data of the known facts and theories were learned and taken from the sites mentioned below as citations. This data is used for the purpose of criticism and comparison with the theory of Brahm.

Sources:

1. Wikipedia contributors. (2025, May 26). Nasadiya Sukta. In *Wikipedia, The Free Encyclopedia*. Retrieved 14:12, May 29, 2025,
   from https://en.wikipedia.org/w/index.php?title=Nasadiya_Sukta&oldid=1292414669

2. Wikipedia contributors. (2025, May 28). Brahman. In *Wikipedia, The Free Encyclopedia*. Retrieved 14:24, May 29, 2025,
   from https://en.wikipedia.org/w/index.php?title=Brahman&oldid=1292666967

3. Wikipedia contributors. (2025, May 29). Para Brahman. In *Wikipedia, The Free Encyclopedia*. Retrieved 14:25, May 29, 2025,
   from https://en.wikipedia.org/w/index.php?title=Para_Brahman&oldid=1292851284

4. https://www.teslasociety.com/tesla_and_swami.htm

5. Wikipedia contributors. (2025, April 28). Aether (classical element). In *Wikipedia, The Free Encyclopedia*. Retrieved 14:47, May 29, 2025,
   from https://en.wikipedia.org/w/index.php?title=Aether_(classical_element)&oldid=1287809885

6. Wikipedia contributors. (2025, April 28). Aether (classical element). In *Wikipedia, The Free Encyclopedia*. Retrieved 14:47, May 29, 2025,
   from https://en.wikipedia.org/w/index.php?title=Aether_(classical_element)&oldid=1287809885

7. Wikipedia contributors. (2025, May 5). Johann II Bernoulli. In *Wikipedia, The Free Encyclopedia*. Retrieved 14:52, May 29, 2025,
   from https://en.wikipedia.org/w/index.php?title=Johann_II_Bernoulli&oldid=1288985121

8. Wikipedia contributors. (2025, May 24). Le Sage's theory of gravitation. In *Wikipedia, The Free Encyclopedia*. Retrieved 14:57, May 29, 2025, from https://en.wikipedia.org/w/index.php?title=Le_Sage%27s_theory_of_gravitation&oldid=1292053290

9. Wikipedia contributors. (2025, May 23). Aether drag hypothesis. In *Wikipedia, The Free Encyclopedia*. Retrieved 14:58, May 29, 2025, from https://en.wikipedia.org/w/index.php?title=Aether_drag_hypothesis&oldid=1291839320

10. Wikipedia contributors. (2025, May 24). Aberration (astronomy). In *Wikipedia, The Free Encyclopedia*. Retrieved 15:16, May 29, 2025, from https://en.wikipedia.org/w/index.php?title=Aberration_(astronomy)&oldid=1291944530

11. Wikipedia contributors. (2025, May 29). Michelson–Morley experiment. In *Wikipedia, The Free Encyclopedia*. Retrieved 15:16, May 29, 2025, from https://en.wikipedia.org/w/index.php?title=Michelson%E2%80%93Morley_experiment&oldid=1292836411

12. https://www.reddit.com/r/QuantumPhysics/comments/1jgj2pd/weekly_famous_quotes_discussion_thread_robert/

13. https://www.reddit.com/r/QuantumPhysics/comments/1jgj2pd/weekly_famous_quotes_discussion_thread_robert/

14. Higgs field. (2025, March 29). *Wikipedia*. Retrieved 15:26, May 29, 2025 from https://simple.wikipedia.org/w/index.php?title=Higgs_field&oldid=10159019.

15. Higgs field. (2025, March 29). *Wikipedia*. Retrieved 15:26, May 29, 2025 from https://simple.wikipedia.org/w/index.php?title=Higgs_field&oldid=10159019.

16. Higgs field. (2025, March 29). *Wikipedia*. Retrieved 15:26, May 29, 2025 from https://simple.wikipedia.org/w/index.php?title=Higgs_field&oldid=10159019.

17. Wikipedia contributors. (2025, May 26). Rømer's determination of the speed of light. In *Wikipedia, The Free Encyclopedia*. Retrieved 13:57, May 29, 2025,

from https://en.wikipedia.org/w/index.php?title=R%C3%B8mer%27s_determination_of_the_speed_of_light&oldid=1292284320

18. Wikipedia contributors. (2025, May 22). Luminiferous aether. In *Wikipedia, The Free Encyclopedia*. Retrieved 14:06, May 29, 2025, from https://en.wikipedia.org/w/index.php?title=Luminiferous_aether&oldid=1291680370

19. Higgs field. (2025, March 29). *Wikipedia*. Retrieved 15:26, May 29, 2025 from https://simple.wikipedia.org/w/index.php?title=Higgs_field&oldid=10159019.

20. Higgs field. (2025, March 29). *Wikipedia*. Retrieved 15:26, May 29, 2025 from https://simple.wikipedia.org/w/index.php?title=Higgs_field&oldid=10159019.

21. Wikipedia contributors. (2025, May 22). Inertia. In *Wikipedia, The Free Encyclopedia*. Retrieved 15:48, May 29, 2025, from https://en.wikipedia.org/w/index.php?title=Inertia&oldid=1291611207

22. Wikipedia contributors. (2025, May 22). Inertia. In *Wikipedia, The Free Encyclopedia*. Retrieved 15:48, May 29, 2025, from https://en.wikipedia.org/w/index.php?title=Inertia&oldid=1291611207

23. Wikipedia contributors. (2025, May 22). Inertia. In *Wikipedia, The Free Encyclopedia*. Retrieved 15:48, May 29, 2025, from https://en.wikipedia.org/w/index.php?title=Inertia&oldid=1291611207

24. Wikipedia contributors. (2025, May 25). Big Bang. In *Wikipedia, The Free Encyclopedia*. Retrieved 15:53, May 29, 2025, from https://en.wikipedia.org/w/index.php?title=Big_Bang&oldid=1292240962

25. Written: 1916 Source: Relativity: The Special and General Theory © 1920 Publisher: Methuen & Co Ltd First Published: December, 1916 Translated: Robert W. Lawson (Authorised translation)

26. Planck scale and plank units: Wikipedia

27. Wikipedia contributors. (2024, December 16). Nicolas Fatio de Duillier. In *Wikipedia, The Free Encyclopedia*. Retrieved 13:04, May 31, 2025, from https://en.wikipedia.org/w/index.php?title=Nicolas_Fatio_de_Duillier&oldid=1263392999

28. Wikipedia contributors. (2025, May 25). Mechanical explanations of gravitation. In *Wikipedia, The Free Encyclopedia*. Retrieved 13:05, May 31, 2025,
from https://en.wikipedia.org/w/index.php?title=Mechanical_explanations_of_gravitation&oldid=1292238158

29. https://www.mathpages.com/home/kmath041/kmath041.htm

30. https://en.m.wikipedia.org/wiki/Planck_units#:~text=synchronization%20of20clocks.-
.Planck%20scale,of%20gravity%20within%20current%20theories

↔→↔

I have learned from the various sources, and it is possible that some known facts mentioned within this book lacks citations. I tried to criticize most of the known scientific theories and Laws of physics, it is not the intention of this book to represent known theory or Data, this book is intended to explain Aether-Brahm theory and its relationship with other theories and Data. History of Aether and kinetic theory of gravitation is mentioned to give credits to the scientists who walked on the way to this theory.

↔→↔

Index